The Transformation of International Agricultural Research and Development

The Transformation of International Agricultural Research and Development

Edited by
J. Lin Compton

Lynne Rienner Publishers • Boulder & London

Published in the United States of America in 1989 by
Lynne Rienner Publishers, Inc
1800 30th Street, Boulder, Colorado 80301

and in the United Kingdom by
Lynne Rienner Publishers, Inc
3 Henrietta Street, Covent Garden, London WC2E 8LU

Library of Congress Cataloging-in-Publication Data
The Transformation of international agricultural research and development

Bibliography: p
Includes index
1. Agriculture—Research—Developing countries
2. Agriculture—Developing countries—Technology transfer
3. Agriculture—Research—United States
4. Agricultural development projects. 5. Agricultural assistance, American. 6. Agricultural extension work—Developing countries. 7. Agricultural experiment stations—United States. 8. Agriculture—Research—International cooperation. I. Compton, J. Lin
S542.3.T72 1989 630'.72 88-29728
ISBN 1-55587-146-1 (alk. paper)

British Cataloguing in Publication Data
A Cataloguing in Publication record for this book is available from the British Library

Printed and bound in the United States of America

The paper used in this publication meets the requirements of the American National Standard for Permanence of Paper for Printed Library Materials Z39.48-1984

Contents

Foreword

> To educate analytical, articulate food and agricultural managers and scientists . . . to meet the food and fiber needs of the United States and the world now and in future decades.
>
> Daniel G. Aldrich, Jr.

With these words, Daniel G. Aldrich summed up his 1985 Justin Smith Morrill Memorial Lecture to representatives of the land-grant universities. These goals and objectives in the education of agricultural scientists have been articulated and demonstrated during the past 100 years in the remarkable history of colleges of agriculture and state agricultural experiment stations (SAES) in the United States.

To celebrate this history and the contributions of state agricultural experiment stations to the institutionalization of agricultural research in developing countries, the Experiment Station Committee on Organization and Policy (ESCOP) commissioned a symposium and this book, *The Transformation of International Agricultural Research and Development*, edited by J. Lin Compton.

The book, with its focus on the transformation of international agricultural research, sketches the influences of the experiment stations in educating foreign students, developing new concepts and research results, and extending or transferring knowledge and technologies to developing nations. The recognition of social, economic, and institutional change brought about over the past century and the potential for a continuing transformation of agriculture and nations is reflected in this and other volumes commissioned by ESCOP. Each volume serves as a benchmark of the past and exploration of the future. The objectives of synthesizing the benefits of agricultural research and education and identifying impacts, especially social and economic, have stimulated the work of the authors and editor.

To undertake these efforts, ESCOP established a Subcommittee on Social Contributions of Agricultural Research. Members included James Halpin (SC); Martha Russell (MN); Noland VanDemark (NY), initial chairman; Lawrence Busch (KY); Ruth Deacon (IA); George Donohue (MN); James Dunn (NY); Walter Huffman (IA); Orville Thompson (CA); and Don Wells (MN).

On behalf of ESCOP and all state agricultural experiment station directors, we extend our appreciation and gratitude to all the authors and the editor of this book for the perspectives, wisdom, and insight they demonstrated in examining the role SAES played in international agricultural development.

This book and others were commissioned as part of the centennial celebration of the passage of the Hatch Act of 1887, in which the federal government authorized the establishment of and continuing financial support of agricultural experiment stations in each state.

The Experiment Station Committee on Organization and Policy is indebted to the several state agricultural experiment stations for granting time and travel support for a number of state supported personnel involved in the committees and task forces that have brought this study to fruition. The support of the Cooperative State Research Service of the United States Department of Agriculture is also gratefully acknowledged.

James J. Zuiches
Associate Dean and Director of Research
Agricultural Research Center
Washington State University
Pullman, Washington

Noland L. VanDemark
Professor Emeritus and Former Director
Cornell University Agricultural Experiment Station
Ithaca, New York

Acknowledgments

This book, which grew out of a larger effort to commemorate the centennial of the establishment of the U.S. State Agricultural Experiment Stations, has involved many persons in addition to the authors of the nine chapters presented. All members of the Subcommittee on Social Contributions of Agricultural Research had a voice in establishing a mandate for this volume. Jim Halpin and Noland VanDemark were especially important voices in our deliberations.

Various members of the faculty at Cornell University were involved in a workshop review of the initial drafts of the chapters. The contributions made by Gene Erickson, Bill Brodie, Njoku Awa, Kathryn March, Harold Capener, Bill Pardee, Norman Uphoff, Ed Oyer, Bob Bruce, Don Schwartz, and Randy Barker to the improvement of our earliest efforts is much appreciated.

The directors of State Agricultural Experiment Stations in California, Florida, Illinois, Iowa, Kansas, Minnesota, New Mexico, and Pennsylvania supported the travel of our authors to the workshop. Our thanks are extended to C.E. Hess, K.R. Tefertiller, B.A. Jones, J.P. Mahlstede, S.E. Leland, R.J. Sauer, and S.H. Smith for their support. The associate administrator of the Cooperative State Research Service (CSRS) of the U.S. Department of Agriculture, C.I. Harris, was instrumental in arranging a grant to cover the technical editing costs of the book. Ellen Bonn of Cornell's Media Services warrants a special vote of thanks for her careful shepherding of the manuscript through the technical editing process. The contributing authors are thankful to all those who gave moral, intellectual, and monetary support to the production of this volume. It is by no means a definitive work. We hope that it will provide a stimulus for the rethinking of scholarship on this important subject and perhaps lead to some progress in the design and conduct of agricultural development programs.

J. Lin Compton

Contributors

Chris O. Andrew, Food and Resource Economics, University of Florida

Virginia Caye, Private Consultant in International Development

R. D. Colle, Communications, Cornell University

J. Lin Compton, Extension Education, University of Wisconsin-Madison

Clyde Eastman, Agricultural Economics, New Mexico State University

Cornelia Butler Flora, Sociology, Kansas State University

Jan L. Flora, Sociology, Kansas State University

James Grieshop, Applied Behavioral Sciences and Cooperative Extension, University of California at Davis

Peter E. Hildebrand, Food and Resource Economics, University of Florida

Everett M. Rogers, Communications, University of California at Berkeley

Vernon W. Ruttan, Agricultural Economics, University of Minnesota

Carolyn Sachs, Rural Sociology, Pennsylvania State University

D. M. Warren, Anthropology, Iowa State University

Robert K. Waugh, Private Consultant in Agricultural Research

The Transformation of International Agricultural Research and Development

Introduction

J. LIN COMPTON

The institutionalizing of agricultural research in developing countries has had a profound impact on food production, the quality of rural life, and the nature of the scientific enterprise in general. In fact, the very character of the process of promoting development is itself being transformed, resulting in a new emphasis on the need to establish generative capabilities and culturally appropriate institutional forms within the developing countries. This emphasis is, at least in part, a reaction to an emerging awareness of previously poorly conceived efforts to transfer institutional forms and agricultural technologies developed in a country whose culture and resources are very different from those of the developing countries. This book assesses this transformation of the nature and functions of agricultural research and development institutions and their programs, describes the major changes, and identifies key trends and issues. It also documents the influence of U.S. initiatives on agricultural development and the relative effectiveness and efficiency of various agricultural research and extension education strategies which have been tried in the developing countries.

Origin and Rationale

In various deliberations that led up to the purpose and design of this book, expressions of concern were made repeatedly about our collective need to learn from a historical accounting of past successes and failures in the large effort to promote world food security. The editor and authors agreed that we would not be able to undertake a series of costly and time-consuming empirical studies. Such a series of studies, while unquestionably worthwhile, would have also required funding that was not available to us at the time.

We decided to proceed and prepare statements that nevertheless would reflect our understanding of past events and current situations and to project our views of future needs and possibilities of Third World agricultural development. The nine chapters represent perspectives and insights borne out of years of scholarship and first-hand practical experience with the subject matter. The analysis and conclusions drawn in the chapters of this book can at best inspire and partially guide later empirical assessments.

The book, then, is a partial assessment of the impact of U.S. experiences and traditions on agriculture in Third World countries and weighs the influence of the institutional models and research and education traditions of the U.S. land-grant university, agricultural experiment station, and Cooperative Extension Service on agricultural institutions, policies, and programs in the countries of Asia, Africa, Latin America, and the Middle East.

Format of the Book

The content of this book covers a broad range of the concerns and issues currently on the agendas for meetings of scientists, administrators, policymakers, and practitioners involved in agricultural programs in the developing countries. In Part 1, Chapters 1 and 2 deal with the context of past efforts in international agricultural development. Chapter 1 explores the history of past policies and programs and assesses the ways in which these have been influenced by political, economic, and social contexts. Chapter 2 is a case study of efforts to transfer a particular technology and related institutional forms from the United States to a developing country. These two chapters together provide a meaningful context for a more specific treatment of different problems and issues.

Part 2 comprises Chapters 3 through 7, which discuss a selection of key problem areas confronting most developing countries today as they proceed to design and implement agricultural development programs. Chapter 3 discusses concerns revolving around the question of how to improve the communication of information among scientists, administrators, educators, and farm families. Chapter 4 explores the questions and issues related to improving the status and role of women in development programs. Chapter 5 assesses the problems and prospects of achieving a greater coherency of effort between research and extension organizations in the developing countries. Chapter 6 is an accounting of the development of extension work in the United States and the successes and failures in efforts to transfer the United States model abroad. Chapter 7 analyzes the nature of farmers indigenous knowledge and the prospects for rationalizing more useful interactions among scientists and farmers in the Third World.

The two chapters in Part 3 focus on the training of scientists and administrators, the functioning of the international agricultural research center network, and the development of farming systems research and extension methodologies, areas in which future technical and financial assistance from the more agriculturally developed countries may have important impact on food production in Third World nations.

Collectively, and with the sequence of their presentation, these nine chapters suggest that a transformation of policies, institutions, and programs of agricultural research and agricultural development has been taking place, albeit not free from the pain of mistakes. We hope that these writings contain lessons that may prove useful in future efforts to improve the art and science of agricultural development.

PART 1

HISTORICAL PERSPECTIVES ON CONTEXT AND CHANGE IN AGRICULTURAL DEVELOPMENT

Studies of the various contexts within which agricultural research and development programs have been carried out provide windows through which observations of research and development, both past and present, take on new or expanded meaning. Thus, the two initial chapters deal with the general questions: What is the nature of the social, political, economic, and technological contexts that have influenced efforts to develop agricultural research and educational institutions in Third World countries? How have larger societal contexts influenced and in many cases actually determined the ultimate outcome of such efforts?

In Chapter 1, Jan Flora and Cornelia Flora suggest that an understanding of the development of agricultural research and extension structures and processes in the United States can facilitate an appreciation of the problems and prospects of developing these types of institutional structures and processes abroad. The U.S. land-grant university and U.S. experiment station have been the creation of particular social, political, and economic forces. In the United States, the demand for increased food production created by a rapid urban and industrial growth, the availability of land for agricultural expansion, and the existence of populist forms of democracy all combined to support the growth of agricultural-oriented universities and experiment stations. Past efforts to develop such institutional structures in the developing countries have been hindered by the lack of funding mechanisms comparable to those established in the United States through the Hatch, Smith-Lever, and Morrill Acts and the failure to recognize basic differences between agriculture in the United States and in developing countries.

The Floras' discussion of the U.S. agrarian transformation encompasses considerations of the role of agricultural mechanization (and cheap labor in the South) in mass production, the importance of transportation and communication networks, the effects of farmers' movements on market prices, the legislative base of institutional developments, the effects of the pressure of food demands from a growing urban population and industrial sector, and the transition from a "labor-saving" to a "land-saving" orientation in innovations. The authors treat several policy-related questions that currently haunt the organizers of national programs: the threat and the promise of bio-technology, the provision of cheap food for urban populations, land tenure and family farming, urban origin of civil servants assigned to rural areas, and farm preservation programs. They also look briefly at the past performance and current role of U.S. land-grant universities in the Third World.

The chapter by Eastman and Grieshop documents four decades of effort to develop potato technology in Peru and promote its diffusion through an institution-building project. A major conclusion reached through this case analysis is that the benefits of scientific advancements are not likely to be widespread in the absence of suitable economic policies and political stability.

CHAPTER 1

An Historical Perspective on Institutional Transfer

CORNELIA BUTLER FLORA
JAN L. FLORA

Only by understanding the institutional transfer that took place within the United States over time can we understand the strengths and weaknesses of transferring our land-grant institutional structure overseas.

The agrarian transformation in the United States began with the Industrial Revolution in the early nineteenth century and snowballed from World War I to the present. It was to a considerable degree a technological transformation. However, that transformation did not occur in isolation. It occurred because of certain social, political, and economic traits in the U.S. system. Agrarian changes were always linked to and usually dependent on the urban-industrial transformation and the institutional structure that engendered it. Only by understanding the relationships between urban demand (for both agricultural products and displaced rural labor) and the growth of agricultural institutions over time can we form realistic expectations for the transference of those institutions to developing countries.

Development of the U.S. Land-Grant Model

By the nineteenth century, the vast majority of U.S. agriculture was market-oriented. The emergence of the concept of the farmer-as-businessman corresponded with the growing subordination of agriculture in the national economy and polity to the needs of the urban industrial system. The expansion westward was fueled simultaneously by industry's need for new supplies and markets and by the drive of unestablished men and women for land and an independent way of making a living. Agricultural research and the land-grant system were outgrowths of these two sets of needs.

During the nineteenth century, the United States Department of Agriculture (USDA) land-grant system played only a marginal role in agricultural politics and technology. (The USDA was not established as a cabinet-level department until 1889.) Land-grant colleges, authorized under the Morrill Act of 1862, were little more than glorified high schools until near the turn of the century (Geiger 1978). This was because rural education prior to that time had been limited to the one-room country schools, which in themselves represented a value of universal education. The land-grant colleges gave rural youth outside New England a chance for the kind of advancement education represented. College, however, for those who were able to leave the farm long enough to go beyond "preparatory" courses to the nineteenth-century land-grant colleges, was not focused on the agricultural enterprise.

The Hatch Act authorized the experiment station system in 1887. In the early meetings called by the commissioner of agriculture to organize the work there was debate over whether the experiment station system should be centralized under the then commissioner of agriculture, or be associated with the land-grant colleges, which were answerable to the state legislatures. The decentralized perspective was accepted by both the land-grant colleges and the commissioner of agriculture. Decentralization allowed the experiment stations to be responsive to local needs. It also ensured that applied rather than basic agricultural research would predominate at the state level for the next century (Knoblauch et al. 1962; Busch and Lacey 1983, 10). The research agenda at the experiment stations was to transform the craft of farming into a science.

The USDA had created its own, more basic, agricultural research system, and made some progress in understanding insects as vectors for disease (Rasmussen 1960, 146), later applying the germ theory of illness, discovered in Europe, to livestock (Rothstein 1978, 124). In 1903, a USDA scientist discovered a serum-virus treatment for hog cholera (Rasmussen 1960, 173-78). Most of the major innovations in the nineteenth century were in agricultural machinery and occurred outside the USDA land-grant system, before the establishment of experiment stations.

The early agricultural research sponsored by the Hatch Act was of a long-term nature with few dramatic breakthroughs. Rothstein characterized its accomplishments in the nineteenth century as not improving the volume of production significantly. While science applied to the agricultural enterprise could forestall certain kinds of disaster, it could not yet make two blades grow instead of one, except by hit-and-miss crossbreeding, or by scouring the world for special cultivars (Rothstein 1978, 124).

Most of the research gains were made by the USDA; the experiment stations had not yet acquired sufficient funding and organization to have a national impact. (An exception was the development by Babcock, a University of Wisconsin scientist, of a successful test for butterfat content in milk, which "rewarded the more efficient dairymen and punished the laggards in the marketplace" [Rothstein 1978, 124; see also Rasmussen 1960, 152-58]).

The subordination of agriculture to the industrial system is also illustrated by the "Country Life" movement (strongest between 1909 and 1919) led by what Danbom (1979) calls urban agrarians and social scientists. (There was only one farmers' representative on President Theodore Roosevelt's Country Life Commission, and few rural people were actually involved in the movement.) Although the two groups differed in their assessment of the positive aspects of rural life, they agreed that agriculture was technically backward and that this was detrimental to urban America, resulting in high food prices for the growing industrial and urban middle classes. The Country Life movement advocated strengthening and transforming agrarian institutions in order to promote scientific agriculture. The consolidation of secondary schools, where vocational agriculture and "enriched" studies could be offered, was promoted along with adult education, which later became the extension service.

The establishment of the extension service through the Smith-Lever Act of 1914 was the result of a number of influences. Seaman Knapp, a scientific farmer, entrepreneur, and president of Iowa State College, established the concept of demonstration farms with a swamplands development scheme in Louisiana in which purchasers were encouraged to plant rice. This established rice as a significant American crop (Mawby, 34-36).

Knapp was hired by the Department of Agriculture in 1902 to improve southern agriculture. Working in Texas with the financial assistance of what later became the Rockefeller Foundation, he used farm demonstration agents to show farmers new cultural practices that would eliminate the boll weevil. The practices were carried out on demonstration plots on the farmers' own land. That approach to eliminating the epidemic helped legitimize the extension concept. Ironically, the American Federation of Labor, the railroads, and various business organizations, such as the National Soil Fertility League, as well as the Country Life movement, were in many regions more supportive of the extension service than farmers, who still believed that science had little to offer agriculture (Danbom 1979, 71-72). Although many established farmers resented the young professionally trained and "citified" agents who came to teach them how to farm, many others eagerly flocked to see exhibits aboard demonstration trains and attended institute lectures and short courses. Farmers and their wives generally wanted to eliminate drudgery and overcome the lack of amenities in their lives (Scott 1970).

Local farm agents were paid with a combination of federal, state, and local government monies, as well as private funds. The agents organized farm bureaus, which resulted in a multiplying effect in the dissemination of information. These bureaus were organized into the national Farm Bureau Federation and provided an organized constituency for the extension, research, and land grant-system in general—particularly vis-à-vis state legislatures. The organization of the farm bureaus, and subsequently county extension councils, brought the more prosperous and politically conservative farmers into the movement for scientific agriculture.

The politically popular argument that farmers would share in the prosperity of scientific agriculture failed to take into account the inelasticity of demand for food crops. Underproduction, rather than overproduction, was generally in the farmers' interest. The cyclical rise and fall of farm prices responded to the interaction of national economic trends and production cycles.

The expansion of the urban population through massive in-migration and the demands of World War I and the immediate postwar period brought about the Golden Age of agriculture for farmers. Although agricultural production increased slowly during that period, it was far outstripped by demand. Prices for farm products rose steeply. Then with the contraction of foreign demand in the 1920s, overproduction brought about a prolonged agricultural recession and depression.

Curiously, the low prices of the 1920s stimulated farmers to try new farming methods and set them on the path of scientific agriculture. Since there were no collective mechanisms to reduce total production and thus increase prices, farmers sought to maximize their individual production and make up for a low price by increasing the number of units. Farmers purchased tractors in record numbers in order to cultivate more land and remain competitive. That decade marked the beginning of the decline in the number of farms, although the urban depression of the 1930s drove people back to the land until the number of U.S. farms peaked in 1935. The economic rationale for mechanization was strong in the 1920s. Still, by 1930, there were fewer than a million tractors on 6 million farms, compared to 4 million automobiles on farms in that year. Improved roads, rural free delivery postal service, and community telephone companies, all subsidized completely or partly by the federal government, broke the isolation of rural areas and opened them to mechanized agriculture.

As Ruttan points out, innovations in U.S. agriculture have been largely-labor saving, i.e., promoting of labor productivity through mechanization (Ruttan 1982, 29-31), which enabled the farmer-businessman to retain profits, especially in areas where labor was relatively unavailable, such as in the Midwest (Pfeffer 1983, 540-562). Only since the advent of World War II has land-saving research (biological improvement for increased yield and gain, fertilization, and irrigation) borne fruit. In this period, land- and labor-saving policies have complemented each other.

The period from World War II to the present has been characterized by rapid adoption of technology. Although most of the technology adopted between 1945 and 1960 was developed before 1945, the higher postwar prices favored its acceptance. During this time growth in agricultural production based on scientific farming was unprecedented in history.

The farm programs of the 1930s and the postwar period placed their emphasis on limiting acreage to avoid agricultural surpluses. In the postwar period farmers responded to that challenge by adopting land-saving techniques, such as genetic improvement, animal confinement, and use of chemical fertilizers, which had previously been the subject of basic research. In hybrids and chemical fertilizers, in particular, private companies were used in the adaptive research and extension process.

The land-grant system has been very successful in ensuring cheap food (see Ruttan, Chapter 8 in this book, for data on returns to agricultural research in the United States). In the "export" era, the system is called upon to expand production to counter a negative balance of trade. In addition to improvement in productivity of plants and animals (and the potential for genetic engineering to speed this process), the land-grant university is also involved in understanding and documenting patterns of international trade, and in some instances in cooperating with the private sector, particularly commodity organizations, to promote foreign sales of agricultural products. As was true in the progressive era, during the Great Depression, and in the period of postwar prosperity and massive migration from farm to city (see Geiger 1978, 32), the client is predominately the large family farm. Only the predominately black land-grant colleges have developed systematic programs aimed at limited-resource U.S. farmers.

The USDA land-grant system is the embodiment of public policy toward agriculture in the United States. It has influenced not only farmers, input suppliers, marketers, and processers, but has also provided the principal model for Americans working overseas in agricultural and rural development. A summary follows of those elements in the model that are relevant (either because they are or are not appropriate) to overseas development. (See Flora 1983 for an expansion of these issues and their implications.)

Food Policy

The introduction of scientific agriculture in the United States initially met considerable resistance from farmers, since its principal objective was to provide cheap food to urban residents. Improving the farmer's welfare, though espoused, was only a secondary objective. In the short term, the farmer's welfare and the cheap-food objective were antagonistic. The welfare of farmers did improve in the long run, but such improvement stemmed in part from reduced competition as some farmers were weeded out—farmers who were late adopters, nonadopters, or who implemented new techniques inappropriately, especially by overcapitalizing.

In developing countries, cheap-food policies have even greater political importance than in the United States. Most countries subsidize staple foods to the consumer. Attempts to raise food prices have often precipitated riots. Almost invariably, low food prices are a lack of incentive for the producer and therefore encourage a continuance of traditional methods tending toward subsistence production. Agricultural development programs that fail to address these policy questions have limited effectiveness.

Appropriate Technology

An expanding urban industrial complex and a rapidly expanding need for industrial labor, especially in the urban service sector, left room for labor-saving techniques in U.S. agriculture without creating massive unemployment. The ability of the larger economy to absorb the noncompetitive farmers in other sectors allowed research and extension efforts to focus on the more successful, larger family farms. While this focus has fostered massive rural-to-urban migration with attendant adjustment problems, as well as the persistence of poverty among limited-resource farmers who refused to migrate, these problems in no way compare in magnitude to those of rural people in developing countries.

With even more rapidly growing populations, access to more labor-saving industrial technology, and peasantries that are more subsistence-oriented than American frontier farmers ever were, today's developing countries require more land-saving approaches than were developed in the United States. The selection process, whereby those less prone to adopt scientific agriculture have to leave farming, is not an acceptable solution in developing countries. New approaches are needed that are appropriate to a variety of farmer circumstances. Extension efforts should be designed to reach a broader spectrum of farmers than did U.S. extension programs, aimed as they were at progressive farmers with adequate resources.

To the degree that the concern is with increasing food production, as it was in the United States, it does no good to provide the larger farmers with access to resources since they are generally not the farmers who produce food for the domestic market. Because of price and other incentives aimed at increasing foreign exchange, large modern farmers are engaged in export agriculture, while food production is generally carried out by limited-resource farmers. The need to reach a clientele that is less formally educated and that is oriented toward traditional methods with fewer resources than nineteenth-century U.S. farmers suggests that the time lines established for research and extension programs in developing countries are much too short. Although it took over a half century for our land-grant system to become firmly established, we have expected major results in Third World countries from a five-year project span.

Democratization of Agrarian Structures

Despite the inequalities created within the agrarian sector of this country, our agricultural history was one of abundant land, comparatively egalitarian land distribution policies, and for all regions of the country except the South and California, an absence of a subordinated racial or ethnic population large enough to form a socially distinct permanent labor force. With the exception of the cotton South, a dualized agriculture in which large plantations produced industrial crops and a subordinate group of small farmers produced subsistence and food crops did not develop in this country.

The land-grant system, once it was accepted by farmers, contributed to the democratization of the agrarian structure. That relative equality within the agrarian sector has meant that farmers value a combination of mental and manual work. One reason the farming lifestyle has been valued in the United States is that the farm family controls the means of production. Farmers generally own their own land, or at least have title to it, even if it's mortgaged. In the United States we see nothing intrinsically wrong with people getting their hands dirty and sweating in the fields, because it is understood that the same people who do the dirty manual work also make decisions about planting, harvesting, purchasing, and marketing. They also keep any profits. Because of shortages of labor in the Northeast and Midwest, U.S. men and women have had to work hard for their food. This practical reality had its forerunner in the yeoman farmer of the eighteenth century and its philosophic underpinnings in the writings of Thomas Jefferson.

How different this is from many countries of the world where land is much less equally distributed. A few landowners have a lot of land, which often is farmed by sharecroppers. Even in places in Africa where land is more equitably divided, the people who make the final decisions about production often live in urban areas and do not do the manual farm work (Clarke 1980, 177-219; Bernstein 1977, 60-73). The idea that getting your hands dirty and sweating is inappropriate for gentlemen and ladies was a class concept either of an indigenous origin or perhaps more often introduced by colonial masters, who were able to hire or coerce the local population to do the manual work, defining it as inappropriate for those in command. Education in such a hierarchical setting is viewed as a tool for getting away from manual work, not for doing it better.

Agrarian Origin of Scientists

As a result of our comparatively egalitarian agrarian social structure, the unity of goals between researchers and farmers seems implicit. In the United States, land-grant researchers understood the objectives and the problems of farming households because both researchers and extension agents—the employees of the land-grant system—came from farming backgrounds (Busch and Lacy 1983). At least in the early days, the researchers and extension agents themselves continued to farm; Seaman Knapp is an example. The farmer's goals had never to be established since they were implicit in the frame of reference and upbringing of the researchers, the extension agents, and the users of research. All shared the same social background and the same experience in practical agriculture.

This shared perspective contrasts sharply with the situation in today's developing countries. Limited-resource farmers often are unable to send their sons and daughters to grade school—much less college or graduate school—to learn agricultural technology. Those who get into the formal agricultural research and extension system tend to be people of urban backgrounds, often from upper-class or elite families. They have never engaged in the hard, physical labor of agricultural production. While training in leading U.S. universities, these developing-country researchers formulate goals according to their scientific interests, not farmers' needs. Their reference group will more likely be made up of fellow scientists rather than limited-resource farmers. Nor will the kind of research undertaken necessarily correspond to the needs of full-time farmers, particularly those marginal farmers who produce most of the food for their countrymen.

Egalitarian Political Base

Over time, a strong constituency for agricultural research and extension has developed in the United States. Although the masses of farmers were indifferent to the legislation establishing the Department of Agriculture and the land-grant research and extension systems, a handful of agricultural scientists, urban agrarians, and businessmen, many with agricultural backgrounds, perceived the importance of a productive agriculture for the growing industrial system. Often, prosperous, white, male farmers were elected to legislatures and to Congress, and expressed their funding needs directly to those bodies. The organization of the bipartisan farm bloc in Congress in 1921 (Rasmussen 1960, 223-26) was rooted in nineteenth-century populism and continued congressional support for farm-related legislation until the 1970s. The farmer constituency came to be organized as pressure groups, first in the form of the Grange, which went through a highly politicized period in the late 1860s and 1870s before it settled down to goals for farmers. Later, farm interests were expressed through the Farm Bureau, which served as the grassroots link to the extension system and continuously promoted the farmer-as-businessman concept among legislative bodies. That promotion generated strong support for agricultural research and extension through the land-grant system.

Operators of large farms in developing countries, who produce industrial agricultural products largely for export, have considerable political power and often are able to obtain favorable prices, export subsidies, and preferential importation of their inputs. However, limited-resource farmers, who produce most of the food for domestic consumption, are often only marginally involved in the political process. Such farmers generally lack organizational presence at the national level. Where they are organized, their principal concerns are with prices and access to resources such as land, credit, and inputs; extension and research are of secondary concern to them. The mandate for research and extension programs for such farmers often comes, instead, from outside donors who are concerned about the macro-problems of food production. Such outsiders from international agencies or the capital city try to understand the farmers; but their understanding is often incomplete, because of the "top-down" nature of the programs.

Values of Family Farming

Finally, the triumph of the farmer-as-businessman approach in the United States has major implications for the relevance of North American expertise to the developing countries. Family farming is characterized by the union of production and household units. Hence, a family farm can never be completely given over to what Weber (1978, 161-64) calls capital accounting—the reduction of all activities to measurement in dollars and cents. He calls such behavior "formal rationality." The family farm can be said to operate under "substantive rationality," in which economic activity may be carried on for reasons other than maximizing profit. The standard for substantive rationality might be the preservation of the farm for the next generation (including conservation practices that are economical only in the long run), a strongly positive valuation on rearing one's children in the country, or providing employment for family members. Such a perspective we call "farming as a way of life." Among limited-resource farmers in developing countries, other substantive goals may be added to those already mentioned: risk reduction by limiting purchases of inputs and balancing production for home consumption with production for the market; diversification of the family's on-farm and off-farm economic activities; and expenditure of resources on ritual social events to ensure the family's integration into a larger community of mutual aid.

As U.S. agriculture has blended with industrialism, farming has taken on business characteristics, albeit with the production and household units usually remaining identical in a formal sense. This transformation has occurred through mechanization that reduces family labor participation, through limited but growing use of hired labor for specialized tasks, and through the introduction of hired managers or supervisors.

Although formal rationality in U.S. agriculture has brought about great increases in food production, it means that land-grant university scientists, administrators, and extensionists have less and less to transfer directly from the United States to developing countries, unless it is to their large-scale, export-oriented sectors. That sector, however, already has access to the latest technology and cultural practices, often through the overseas presence of multinational agribusiness firms. Then assistance to that sector is to a considerable degree superfluous.

This does not mean that land-grant personnel cannot make any contribution to improving food production and rural welfare in developing countries. With a growing sophistication and access to resources, the land-grant system now has the institutional capability to work on new approaches and knowledge appropriate for conditions in developing countries. There are still two major barriers, however, to effectively transferring land-grant expertise to developing countries:

1. The unique funding mechanism instituted through the Hatch and Smith-Lever Acts, which has been so effective in bringing technical and administrative innovations to U.S. agriculture, poses a problem to the effective translation of those innovations overseas. Land-grant extension services and experiment stations are quite dependent on state legislatures for funding. It is difficult to demonstrate that overseas work could directly benefit any one state. Indeed, if the overseas work were to increase production of a commodity that competes with domestic produce, as is often the case when local scientists are familiar with their own in-state commodities, there can be legislative opposition to it. Although funding for overseas projects comes from federal or private sources, legislators and land-grant administrators often see them as taking human resources away from the "real" mission of the land-grant university. Creative administrators, such as at the University of Florida and Cornell University, have been able to convince state legislators of the domestic value of overseas projects, and they have actually obtained modest state appropriations for such activities.

2. If the differences in agricultural structures and farmer rationality in such places as between Nebraska and Colombia or New York and the Philippines are not clearly understood by the land-grant scientists, administrators, and extensionists, their impact overseas will be nil or even harmful. The more clearly those differences can be explained to land-grant personnel, the more effective they can be in their overseas work.

The next section of this chapter examines the history of the land-grant system overseas both directly (through contracts in developing countries) and indirectly (through the implicit or explicit following of land-grant approaches).

The Land-Grant System Overseas

Massive and official attempts by the land-grant universities to replicate the advantages of their unique research, teaching, and extension systems overseas awaited the Point IV Program, which followed the Second World War. Nevertheless, important links from the land-grant colleges were already forming through institutional ties or faculty consultations and student training.

The early involvement of U.S. land-grant universities in overseas agricultural development was idiosyncratic, having grown and institutionalized out of personal contacts. Examples of such programs would be the early work of Massachusetts State College at Hokkaido, Japan, in 1876, or the appointment of a Michigan State University faculty member as president of Brazil's first school of agriculture in 1908, or Cornell University's link with agriculturalists at the University of Nanking in China. From 1924 to 1931, Cornell agricultural experts worked in China, and Chinese students were trained at Cornell in genetics and plant breeding. The joint goal was to increase China's agricultural production and available food supply.

The U.S. desire to maintain and generate markets for itself, particularly in Latin America, was heightened by U.S. foreign policy interests on the eve of World War II. The natural trading partners for Argentina, Brazil, and Mexico were England and Germany, and the United States feared Nazi infiltration in these areas. During World War II, the Office of Foreign Agricultural Relations (OFAR) of the USDA established experiment stations in Latin America. Further, Nelson Rockefeller and his Office for Inter-American Affairs staffed agricultural development work in Latin America with USDA and land-grant college professionals.

In 1941 the Department of State, in cooperation with the USDA-OFAR, commissioned prominent rural sociologists from land-grant universities to make systematic studies of the rural sectors of Brazil, Argentina, Mexico, and, in 1945, Cuba. The three initial studies were of the largest countries in Latin America—and the ones most likely to take an independent foreign policy line. Later technical cooperation activities, such as Point IV and the U.S. Agency for International Development (USAID), were guided by the results of those studies (Smith 1947), which implicitly tended to compare and contrast the existing agrarian structures in the study countries with the ideal of the medium-sized family farm in the United States. In fact, the rural sociologists who went to Latin America were also critical of any structure in U.S. agriculture that mitigated against the survival of the family farm.

In the 1940s, the Colombian government requested a $10 million loan from the United States to relieve the general state of social and economic crisis in Colombian agriculture. The Colombian government asked for outside advice concerning a basic institutional structure—land tenure—to counteract the immobility of the second López administration (1942-1945), which faced concerted landowner opposition to the land reforms of 1936 (Cronshaw 1982, 99).

T. Lynn Smith, of Louisiana State University, advised institutionalizing the land-grant university's ideal client in Colombia. He recommended that all efforts should be made to strengthen and increase an agricultural middle class. Those efforts would be based on the opening of new lands by colonization, however, rather than by directly challenging the existing landholding patterns in established agricultural areas. Smith saw that building a participating, viable family farm population would take at least twenty-five years before substantial progress could be made (Smith 1947, 129-140). The middle-class farmer that the policy hoped to foster combined three basic economic roles: capitalist or entrepreneur, manager, and laborer (Cronshaw 1982, 102). That the evolution from the dual social structure of peasant and *latifundista* would be difficult was recognized by Smith, who was dismayed by the Colombian elite's disdain for manual work. Nevertheless, he viewed the middle-class family farm as the necessary cornerstone of democracy, requiring a relatively high level of education, as well as application of the techniques of scientific agriculture.

In 1941 the Rockefeller Foundation began a program of agricultural development in Mexico that attempted to use agricultural technology and expertise acquired in a developing country setting (Mosher 1957). The Rockefeller Foundation had been a key to expanding Seaman Knapp's farm demonstration model in non-boll-weevil-infested areas of the southern United States between 1906 and 1919 through helping to fund the General Education Board (Fosdick 1962). This experience convinced the Foundation of the validity of the extension component of the land-grant model.

Thus, when the Rockefeller Foundation began to evaluate Mexico as a site for developing solutions, they turned to key agricultural scientists at major U.S. land-grant institutions. The infusion of Rockefeller dollars was to transform Mexican agricultural science from a laboratory exercise into field experimentation.

While efforts at selecting and breeding corn were relatively unsuccessful in terms of increased production, the wheat program, led by Norman Borlaug, resulted in widespread adoption and large production increases. It can be argued that the difference in the impact of the two programs was due less to research results than to differences in the users of the products. Wheat farmers in Mexico were more like U.S. farmers than were the Mexican corn farmers. The wheat farmers were commercial farmers, while the corn farmers were subsistence farmers (Myren 1969).

The Rockefeller program put great emphasis on training agricultural scientists to develop resistant cultivars as opposed to creating delivery systems for the subsistence sector. Since there was no indigenous farmer demand for scientific education, training at the highest levels was much more likely to take hold than grassroots training.

The Office of Special Studies was the Mexican structure used to institutionalize the concept of the experiment station, and the success with which nationals were trained there led to the training of agriculturalists from other countries and to international agricultural conferences. These trained scientists were successful in creating new cultivars adaptable to Mexican commercial farming. Like the U.S. land-grant system it emulated, the Rockefeller-funded programs and the Office of Special Studies did not specifically analyze subsistence farming in Mexico nor produce or diffuse technology for it during the 1941-1953 period of institution building.

During World War II another thread of involvement by U.S. land-grant universities in agricultural development overseas was set in motion. Universities such as Cornell were deeply involved in the U.S. Army's specialized training program, which trained language and area specialists for the armed services. This applied language training put heavy emphasis for the first time on spoken language ability in non-Western as well as Western languages (*Education and World Affairs*, 1965, 183). Such language and area knowledge, as well as government links, were crucial for the success of agricultural development programs.

The United States did not attempt to mobilize a major foreign economic assistance program until near the end of the war. That mobilization attempted to block communist political and military initiatives and to prevent Russia from gaining ground in a war-ravaged Europe. The Marshall Plan, with its large capital investments in Europe's productive infrastructure, was the initial bilateral commitment to development assistance. In current dollars, the investments were the equivalent of $45 billion (Morss and Morss 1982, 19).

It is instructive to look at the Marshall Plan as it contrasts with aid to developing countries. To rebuild Western Europe and Japan, capital and construction were needed, not technical assistance. Institutional structures that favored production had been in place, and were only temporarily interrupted by the war. A highly educated population was able to provide the management and skilled labor necessary for moving from reconstruction immediately into production. Market channels had merely to be revitalized, not invented. Reconstruction was carried out by the private sector, with financing from the public sector.

The immediate and striking success of the Marshall Plan in Europe and Japan and the heightened fear of communist influence in developing countries inspired U.S. planners to attempt a similar operation in the Third World. The Point IV Program, announced by President Harry Truman in his inaugural address January 20, 1949, laid out the rationale for the plan that was to provide economic and technical assistance to help people in developing countries produce more.

By this time, the land-grant colleges were alert to their interests as public sector institutions that could provide overseas technical assistance. Within hours of his speech, the president received a call from John A. Hannah, president of the American Association of Land-Grant Colleges and State Universities, expressing a desire to cooperate in "making Point IV effective" (*Education and World Affairs,* 1965, 49). The academic resources the government had called on during the war were offered to implement national policy in international affairs.

In 1950 Congress authorized the Point IV Program with the Act for International Development. The act created a Technical Cooperation Administration, which was to administer technical assistance to selected countries.

Initial development efforts were based on the twin premises of massive investments in infrastructure and the need to encourage democratic institutions through popular participation. This took the form of community development programs to increase local participation in decisionmaking and efforts by extension to reach operators of small farms. Technology transfer and local political participation were seen as fundamental to economic development and the maintenance of political freedom (i.e., keeping communism at bay).

Economic investments in production could not be realized in the educational and institutional vacuum present in most developing countries. Private foundations, including Ford, Rockefeller, Kellogg, and Carnegie, early recognized the contradictions of trying to apply new technologies to institutionally barren situations. Thus, the major foundations made their initial grants to U.S. land-grant institutions, which in turn would study the local conditions in different parts of the world. The universities were charged with devising practical suggestions for establishing institutions that would do the research and teaching necessary for an agricultural development program—including the training of agricultural extension agents. Early grants from these private foundations to the University of Illinois in India, Michigan State University in Pakistan, the University of Wisconsin in Indonesia, and Cornell in Peru paved the way for larger U.S. government contracts from the Technical Cooperation Administration, which became the U.S. International Cooperation Administration (ICA) and later the Agency for International Development (USAID). The methodology for direct U.S. government institution-building projects around the world evolved from these first initiatives. Further, the community development effort was primarily funded by private foundations and required long-term investment. Money from public sources was expected to produce immediate, measurable results—a demand which meant that few ICA/USAID funds went for community development *or* agricultural research.

Despite the tendency to depict 1950 to 1965 as the community phase of international economic development, it is important to realize the dominance of large investments in institution building, which involved the technical training of national elites and large infrastructural investments. The land-grant institutions were involved only in community development and institution building. Only the latter survived as part of United States international development strategy, in part because of its complementarity with the investment strategy. The institutions would train the nationals of developing countries to use the new technological resources made available through international investment.

The initial model of economic development that directed U.S. post-World War II efforts has been referred to as the resource constraint model. Walt Rostow was the best known proponent of this model. The fifteen years that followed World War II were typified by efforts to increase national, including agricultural, output and by investments to increase capital stock, which included agricultural technology. Morss and Morss refer to this as the "big lever" approach to development, characterized by the following:

1. A focus on increasing aggregate production
2. Macro economic planning to identify constraints to increasing aggregate production
3. A view that output growth was primarily constrained by inadequate technology investment
4. Foreign assistance to eliminate investment shortages

5. Foreign assistance not tied to specific project activities, but directed at: the use of Western equipment and technologies, including community development and traditional agricultural extension techniques; large scale infrastructural investments; education through technical assistance and foreign schooling for students from developing nations; and institution building (Morss and Morss 1982, 22-23)

There was a definite conflict between the economic resource constraint initiative, which favored advantaged classes and a trickle-down approach, and the political participation initiative, which tended to be more concerned about distributional issues. It became convenient to abandon the efforts aimed at increasing participation. The rationale for abandonment was not to note the inconvenience and political sensitivity of this approach, but to label participation a failure (Holdcroft 1978).

It was easy to dismiss the community development work as a failure (ignoring T. Lynn Smith's warning of the long time frame necessary to change social institutions at a grassroots level) and to concentrate on educating developing nations to see the value of new technologies, and provide them with the financial assistance to purchase those technologies. Land-grant institutions were considerably more comfortable training people to see the value of new technologies than training them for democratic participation. The passing of community development from fashionable acceptance was little mourned by land-grant universities doing international work. They were busy replicating themselves with massive investments and personnel transfers in such places as the Philippines, India, Pakistan, Nigeria, Brazil, and Colombia.

In 1962, USAID was created, bringing together a number of previously dispersed foreign aid programs: the International Cooperation Administration; the Development Loan Fund; the Food for Peace (P.L.480) program; and local currency-lending activities of the Export-Import Bank. The new organization was perceived as a permanent agency, unlike its predecessors which had been created to deal with particular international problems (Tendler 1975, 15).

The Alliance for Progress was established in 1961 for a particular purpose and for a limited period. It was initiated in direct response to the Cuban revolution of 1959. It recognized that some of the constraints to progress in the hemisphere stemmed from more than a lack of resources and knowledge. Land reform, an issue Smith was loath to address in the 1940s, became an issue once again in the 1960s.

The concern for land reform among liberal Latin American leaders was less a concern for social justice and more a desire for productivity and a way to deflect peasant agitation for land (Barsky and Cosse 1981). From the U.S. perspective, foreign aid was to become a lever for convincing governments to make policy changes, including land reform, needed to diminish a perceived communist threat. While U.S. land-grant institutions were engaged in setting up research and extension services on a massive scale in Latin America, the thrust of agricultural development in Latin America in the 1960s was toward a more productive agriculture, capable of producing for export. The major landowners were fearful of expropriation if their land continued to be exploited in the traditional extensive manner, so they began to intensify production, investing in the capital improvements and technological inputs necessary to produce such crops as sugar and cotton for the world market. The newly created national research institutions emphasized the crops grown by these modernizing operators, for it was the large farmers who had political influence over the state.

The precise way in which technology transfer occurred with respect to any particular crop in a particular country depended on the type of productive unit, as well as whether the crop was domestically consumed (which militated toward low prices), or principally for export, and the availability of internationally generated technology, as well as the nature of the government itself. In general, new technology was adopted (1) when crops or livestock were the only important enterprise on the farm (thus facilitating organization of politically important commodity guilds); (2) where, through stagnation of production or expansion of demand, the price of a staple food product had risen substantially in price, in which case the demand for increased production came from a group other than the producers; (3) where technology for substantial increase in production was already available internationally, as in the case of high-yielding varieties of rice; and (4) where the state was sufficiently strong to organize the needed research. Technology generation and/or transfer is organized by the public sector, or jointly by the public sector and a commodity guild (as in the case of rice modernization in Colombia). In a number of recent instances, especially in Colombia, farmers have hired their own technical production experts through their commodity guilds and, in the case of the sugar guild, have established their own experiment stations (Piniero and Trigo 1983).

With this privatization of the research system for large-scale commercial agriculture and with temporary pressure on governments to serve peasant farmers as well, brave new research and extension establishments found themselves, by the end of the 1960s, without a constituency. This was the case with the Instituto Agropecaurio Colombiano (ICA) funded by USAID and the Rockefeller Foundation and implemented by a large and dynamic mission from the University of Nebraska. By default, these establishments were left to produce and deliver agricultural technology for the limited-resource farmer. Limited economic resources meant limited political resources. Although peasants were being organized at that time in Colombia by the liberal Lleras administration, their demands were less for technology and extension than for land and improved prices. In many countries, the research and extension institutions had no political backing and systematically lost funding in the 1970s when foreign funding ceased. Trigo, Piniero, and Sabato suggest that the centralized Latin American research institutes lacked "the federalized organization of the land-grant system, where the regional decentralization of its institutions facilitated a strong socio-political linkage with the farmers in each zone" (Trigo et al. 1983, 135). Even within the same region, the social heterogeneity of the farm population is much greater in most Latin American countries than in the United States (ibid.).

By the 1970s, foundation funding was increasingly going to the international agricultural research centers, and it was expected that the cadres of agricultural scientists they trained would return to their domestic institutions. By 1980, few of those trained were to be found in the research and extension institutes in Colombia, Argentina, Peru, or Venezuela (Ardila et al. 1981). The domestic public administrative structures were incapable of retaining these expensively trained individuals.

In contrast, the institutional structures in Mexico and Brazil proved conducive for agricultural scientists, and relatively strong institutions remained into the decade of the 1980s. Although wages were low in the countries that did not maintain their researchers and extension agents, the individuals surveyed stated that working conditions, not salary, were paramount in their decisions to seek employment elsewhere. Institutionalization had not gone beyond the bounds of the academy.

The attempts during the 1960s to establish land-grant institutions around the world achieved only partial success. In countries like India, the national government officially called upon the U.S. government to help set up agricultural universities aimed at training, research, and extension, and ten years of U.S. presence and funding did result in the establishment of a number of strong universities (Read 1974). But the fact that the universities were often set up as a compromise of U.S. scientists with local political interests led to strange configurations. For example, the seed farm at Patnagar, Uttar Pradesh, was set up by the University of Illinois, but employed large numbers of migrant laborers and produced huge profits (Brass 1983, 103-161). A consistent problem for institution building in developing countries comes from a natural lack of awareness among expatriate staff of all possible ramifications in each action, particularly the implications of their activities for the less powerful clients, urban as well as rural.

If the 1960s was the decade of building universities and extension systems with the land-grant model, the 1970s was the decade of building roads and other infrastructure, with private contractors providing goods rather than services. International development in the 1970s was characterized by the slogan "reaching the poorest of the poor." It was reasoned that they could not be reached if there were no roads, and no decent water to drink on arrival (Nicholson 1979, 225). Although agricultural development and Integrated Rural Development Projects (IRDPs) established during the 1970s continued to be important to extending the Green Revolution to farmers with fertilizer and water, USAID substantially shifted its funding focus away from U.S. land-grant institutions in the United States and their counterparts overseas. Once again, quick results were sought over potential long-term gains.

The new directions initiated by the concern for the poorest of the poor and the redirection toward infrastructure with their concomitant and huge international debts resulted in a substantial drop in USAID-financed contracts and grants for technical assistance to host countries. From $180 million in both 1970 and 1971, aid dropped to $46 million in 1976—the nadir of foreign assistance channeled through U.S. universities in recent years.

Declining foreign assistance activity by U.S. universities in the mid-1970s (less than half the contracts in place in 1970) corresponded with both a decline in total USAID funding and with the redirection of development activities. At that time USAID clearly felt that U.S. land-grant institutions could do little to directly improve the lives of the poorest people in the developing world. This position was seen by the universities, however, as an overly narrow interpretation of the mandate. Land-grant universities were able to mobilize their considerable political clout to argue that the emphasis on capital transfer over institution building was inappropriate. Stressing the continuing world food problem, the universities presented themselves as uniquely able to combat famine and reduce hunger in the developing world by applying the research-teaching-extension model to increase productivity.

The result was the passage in 1974 of Title XII of the Foreign Assistance Act. Under Title XII, activities were to

1. Be directly related to the food and agricultural needs of developing countries
2. Be carried out within the developing countries
3. Be adapted to local circumstances
4. Provide for the most effective interrelationship between research, education, and extension in promoting agricultural development in developing countries
5. Emphasize the improvement of local systems for delivering the best available knowledge to the operators of small farms in such countries

The bill also authorized the president to strengthen the capacity of U.S. universities to work overseas. In addition to the reinclusion of U.S. universities in international development work related to food and agriculture, Title XII provided program support for international agricultural research centers. In effect, Title XII gave foreign aid a domestic constituence—the land-grant system—which Tendler (1975, 38-39) argued it had lacked.

A reorientation of U.S. universities would be needed if they were to truly meet the needs of the poorest of the poor—or even limited-resource farmers. In the United States the land-grant system had been aimed, with certain exceptions, at increasing the productivity of farmers with access to resources, not the marginal farmers. Furthermore, when dealing with the latter, it was clear to many people that a simple transfer of technology was totally inappropriate, and could not be done even if it were somehow the best thing to do. The method, not the content, institution, or product, had to be the subject of the new development thrust. And this thrust meant a reorientation. The "easy" productivity gains by the introduction of "miracle" grain varieties in high-fertility conditions had been made. Now what was left were the tough productivity problems—depleted soils in areas of irregular rainfall tilled by peasants with little education, little access to cash, and conflicting demands from the multiple survival strategies of livestock, crop production, and wage labor. Institutional transfer had to begin with making U.S. universities aware of the different conditions under which research and extension would take place. The hope was to forge a partnership between USAID and land-grant universities (United States Comptroller General 1981).

There have been a number of centrally funded efforts by USAID to help U.S. universities retool for the difficult task of serving the limited-resource farmer. These include Strengthening Grants, Cooperative Reseach Support Projects, and Support Projects. The land-grant institutions are encouraged to link with the international research centers. Strengthening Grants give designated academic institutions (identified as Title XII) the opportunity to receive grant funds to strengthen their faculty and institutional skills in working overseas. A number of these grants have become Memorandums of Understanding between USDA and individual colleges or joint 1862 and 1890 land-grant institutions and help to maintain overseas development capacity. The Cooperative Research Support Projects (CRSPs) link U.S. scientists on a commodity basis to their counterparts in developing countries to jointly solve production problems relevant to limited-resource farmers. The Farming Systems Support Project is helping to sensitize scientists at U.S. land-grant institutions to the very different constraints faced by farmers in developing countries. The project also supports bilateral contacts between USAID and land-grant universities and links such universities directly to national research organizations.

The task of reaching the limited-resource farmers who produce the bulk of the domestic food consumed in developing countries is a difficult, long-term one. Funders demand results. For universities this means producing quantifiable indicators for USAID. For USAID, it means producing impressive results in response to congressional inquiry. Thus, there has been a falling away from the goal of reaching the poorest of the poor. This was already acknowledged in the Integrated Rural Development Projects, which tended to place a minimum as well as maximum size on the first targeted farm and to focus on areas of relatively promising agronomic potential.

The lexicon of the "New Orthodoxy" in foreign assistance is to stress "food self-reliance" over "food self-sufficiency." Developing countries may stress production for export if exports produce foreign exchange to buy food grains from the United States. The goal is to reach the market-responsive farmer, rather than the limited-resource farmer. This private sector, market repossession method has yet to address the variety of structures mediating between many farmers (particularly in mixed farming and herding systems) and the market, particularly in Africa. The farming systems approach to research and extension has the potential to meet the needs of complex farming systems, but its iterative nature requires time to yield results—just as input from local farmers was gradually included in research and extension programs in the United States.

Conclusions

In Asia and Latin America, institutions have been built—or at least the centers for training, research, and extension have been established;

and an adequate number of trained agricultural scientists would be available to staff them if the larger institutional setting were more conducive. In these countries, the human capital has been generated to a large degree. What is lacking are the political mechanisms that would allow farmers to have a direct, decentralized input into the institutions and to influence those institutions to address their needs in research and extension.

In the development community, it is generally recognized that large investments in human and institutional capital still need to be made in Africa. Training that would bring African students to U.S. land-grant institutions for degree programs will likely be stressed again. Although such training has greater potential than digging wells that tend to fall into disrepair as quickly as new ones are built, there is also the danger that what will be institutionalized is not the method, but the content. Care must be taken to make students aware that they are not to go back to recreate a U.S. experiment station, complete with monoculture and giant machinery. Like the agricultural scientists at the U.S. stations, they must work with their own nationals—the farmers they hope to serve—to create and deliver technologies that meet local needs.

But even if the research and extension units are finely tuned to local needs, they will still require national political support to serve the limited-resource farmer—or whatever client group is defined. Otherwise, the human capital will dissipate. The newly created agricultural institutions must be able to use Western technology selectively, making choices about what will fit local conditions and what must be reworked to conform to a labor-surplus, capital-short, small farm system (Molnar and Clonts 1983, 7). Further, they will have to do it within a shorter time frame and with less-stable funding than in the early decentralized years of the land-grant system in the United States.

The U.S. land-grant system is, in reality, a complex mix of 1862 and 1890 institutions. Each provides training, research, and extension, according to the unique needs of its defined constituency within the relatively small geographic area it serves. Each has had stable funding, much of it generated locally to ensure responsiveness. While those who were most vocal were the operators of larger farms, there was still a constant legitimation process between the servant and the served. Such accountability is the crucial element of institutionalization and it is difficult to impose from the outside.

References

Ardila, Jorge, Eduardo Trigo, and Martin Piniero. "Human Resources in Agricultural Research: Three Cases in Latin America." PROTAAL Document No. 50. IICA, San José Costa Rica, n. d.

Barsky, Osvaldo, and Gustavo Cosse. "Technología y Cambio Social: Las Haciendas Lecheras del Ecuador." Quito: Facultad Latinoamericana de Ciencias Sociales (FLACSO), 1981.

Bernstein, H. "Notes on Capital and Peasantry." *Review of African Political Economy* 10(1977):60-73.

Brass, Paul R. "Institutional Transfer of Technology: The Land-Grant Model and the Agricultural University at Patnagar." In *Science, Politics, and the Agricultural Revolution in Asia*, Robert S. Anderson, et al., eds. Boulder, Colo.: Westview Press, 1982, 103-163.

Busch, Lawrence, and William B. Lacy. *Science, Agriculture, and the Politics of Research*. Boulder, Colo.: Westview Press, 1983.

Campbell, Christina McFadyen. *The Farm Bureau and the New Deal*. Urbana: University of Illinois Press, 1962.

Clarke, Julian. "Peasantization and Land Holding: A Nigerian Case Study." In *Peasants in Africa: Historical and Contemporary Perspectives*, Martin A. Klein, ed. Beverly Hills: Sage, 1980, 177-219.

Cronshaw, Francine. "Exporting Ideology: T. Lynn Smith in Columbia." *North South* 13(1982):95-110.

Danbom, David B. *The Resisted Revolution: Urban America and the Industrialization of Agriculture, 1900-1930*. Ames: Iowa State University Press, 1979.

Education and World Affairs. The University Looks Abroad: Approaches to World Affairs at Six American Universities. New York: Walker Publishers, 1965.

Flora, Cornelia Butler. "Farming-Systems Research and the Land-Grant System: Transferring Assumptions Elsewhere." *The Rural Sociologist* 3(July 1983):220-28.

Fosdick, Raymond B. *Adventures in Giving: The Story of the General Education Board*. New York: Harper and Row, 1962.

Geiger, Louis G. "The Morrill Act and Its Interpretation Through Practice." In *Outreach Programs of the Land-grant University: Which Publics Should They Serve?*, Jan L. Flora and Jim Converse, eds. Manhattan: Kansas Agricultural Experiment Station, 1978.

Holdcroft, Lane E. "The Rise and Fall of Community Development in Developing Countries, 1950-1965: A Critical Analysis and an Annotated Bibliography." MSU Rural Development Paper #2, Department of Agricultural Economics, East Lansing, Michigan State University, 1978.

Johnson, Twig, et al. "The Impact of PL 480 Title I in Peru: Food Aid As An Effective Development Resource." AID Project Impact Evaluation #47. Washington, D.C.: U.S. Agency for International Development, 1983.

Knoblauch, H. C., E. M. Law, and W. P. Meyer. "State Agricultural Experiment Station: History of Research Policy and Procedure." Miscellaneous Publication No. 904, U.S. Department of Agriculture.

U.S. Government Printing Office, 1962.

Mawby, Russell G. "Agricultural Scotoma: A Limiting Vision of the Future." *Gleanings* (December): 32-45.

Molnar, Joseph J., and Howard A. Clonts, eds. *Transferring Food Production Technology to Developing Nations: Economic and Social Dimensions.* Boulder, Colo.: Westview Press, 1983.

Morss, Elliott R., and Victoria A. Morss. *U.S. Foreign Aid: An Assessment of New and Traditional Development Strategies.* Boulder, Colo.: Westview Press, 1982.

Mosher, Arthur A. *Technical Cooperation in Latin America.* Chicago: University of Chicago Press, 1957.

Myren, Delbert. "The Rockefeller Program in Corn and Wheat in Mexico." In *Subsistence Agriculture and Economic Development*, Clifton Wharton, ed. Chicago: Aldine Publishing, 1969.

Nicholson, Norman K. "The Congressional Mandate for Aid and Rural Development Strategies." In *The Role of U.S. Agriculture in Foreign Policy*, Richard M. Fraenkel, et al., eds. New York: Praeger, 1979.

Pfeffer, Max. "Social Origins of Three Systems of Farm Production in the United States." *Rural Sociology* 48(1983):540-62.

Piniero, Martin, and Eduardo Trigo. "Social Articulation and Technical Change." In *Technical Change and Social Conflict in Agriculture: Latin American Perspective*, E. Trigo and M. Piniero, eds. Boulder, Colo.: Westview Press, 1983.

Rasmussen, Wayne D., ed. *Readings in the History of American Agriculture.* Urbana: University of Illinois Press, 1960.

Read, Hadley. "Partners with India: Building Agricultural Universities." College of Agriculture, University of Illinois at Urbana-Champaign, 1974.

Rostow, Walter. *The Steps of Economic Growth.* Cambridge: Cambridge University Press, 1964.

Rothstein, Morton. "Crisis and Change in a Century of American Agricultural Research." In *Outreach Programs*, Cornelia Butler Flora and Jim Converse, eds. Manhattan: Kansas Agricultural Experiment Station, 1978.

Ruttan, Vernon W. *Agricultural Research Policy.* Minneapolis: University of Minnesota Press, 1982.

Scott, Roy V. *The Reluctant Farmer: The Rise of Agricultural Extension to 1914.* Urbana: University of Illinois Press, 1970.

Smith, T. Lynn. "Colonization and Settlement in Colombia." *Rural Sociology* 12(1947):129-40.

Tendler, Judith. *Inside Foreign Aid.* Baltimore: Johns Hopkins University Press, 1975.

Trigo, Eduardo, Martin Piniero, and Jorge Sabato. "Technology as a Social Issue: Agricultural Research Organization in Latin America." In *Technical Change and Social Conflict in Agriculture: Latin American Perspective*, E. Trigo and M. Piniero, eds. Boulder, Colo.: Westview Press, 1983.

United States Comptroller General. "AID and Universities Have Yet to Forge an Effective Partnership to Combat World Food Problems." Report to Congress. U.S. General Accounting Office, Washington, D.C., October 16, 1981.

Weber, Max. *Economy and Society*, Vol. 1, Guenther Roth and Claus Wittich, eds. Berkeley: University of California Press, 1978.

CHAPTER 2

Technology Development and Diffusion: Potatoes in Peru[1]

CLYDE EASTMAN
JAMES GRIESHOP

Peru provides a particularly appropriate example of foreign adaptation of the U.S. land-grant model. Peru has had a long and generally friendly relationship with the United States and has sent agricultural scientists to this country for training for well over half a century. Since 1942, there has been some form of active U.S. agricultural technical assistance in Peru involving research or extension programs. From the mid-1950s to the mid-1970s, two major land-grant universities provided much of this assistance—Iowa State and North Carolina. The North Carolina State University Mission focused its efforts on building up the research and extension capacity of the Ministry of Agriculture and the instructional capability of the National Agrarian University (*Universidad Nacional Agraria,* or *UNA*). The North Carolina Mission was closed at the end of 1972; the Iowa Mission continued at a reduced level for several more years. In 1971, the International Potato Center (*Centro Internacional de la Papa,* or *CIP*) was established at La Molina and has continued specialized assistance not only to Peru, but worldwide.

Thus, for a period exceeding four decades there has been an organized and sustained U.S. effort to develop technology overseas and to promote its diffusion. In what ways has Peru benefited from the scores of research projects, hundreds of scientist years, and millions of dollars invested? What have been the primary results of this exemplary effort to improve institutional capacity modeled on the U.S. land-grant system? There was never a sustained effort to build an integrated replica of a land-grant institution in Peru. However, anyone familiar with the U.S. model would recognize its strong influence on the organization and programs of the extension service and on the research arm of Peru's Ministry of Agriculture. The curriculum of UNA was Americanized during the 1960s, although the teaching function remained separated from extension and from research as well.

The above sketch suggests the history and outgrowth of efforts in Peru. Whether or not the Peruvian case is typical of the hundreds of other land-grant exports to foreign universities, agricultural ministries, and research stations throughout the world is, in reality, a moot point. Simply stated, there is little that is "typical" in the worldwide context of diverse cultures and political systems. Nevertheless, the Peruvian case, which represents years of effort and substantial resources, merits examination. This case can provide an understanding of both the benefits and limitations of efforts to transfer research and institutional capabilities.

We will first describe the evolution of the technical capability of the Peruvian Ministry of Agriculture and the National Agrarian University from the first decade of this century to the present. We will then describe the program of the North Carolina Mission and briefly examine other important influences on Peruvian agricultural institutions. Next we will describe the institutional effects on indigenous agriculture, and then examine potato and other agricultural production during the period in question.

Finally, we will interpret the findings and offer observations for future institution-building efforts. Throughout these discussions, the major context and reference point is potato research and the provision of foreign assistance related to that effort. Potatoes received as much or more technical assistance than any other commodity. The focus on one primary food commodity offered more depth of analysis, yet still permitted a manageable undertaking. Less detailed data are also included on other major commodities to demonstrate that potatoes were not a unique case in Peruvian agriculture.

Research from 1901 to 1943[2]

Agricultural research in Peru began in 1901 with the founding of a series of institutions. In that year, the Cotton and Rice Experiment Station and the Wine-viniculture Laboratory were founded. In 1902, the National School of Agriculture was founded at La Molina, followed in 1906 by the Sugarcane Experiment Station. Other one-crop stations were subsequently established in other locations in the country, specifically at Moquegua, Lambayeque, Iquitos, and Ica. In January 1912, the Central Agronomy Station was created and passed in 1919 to the Technical Division of Agriculture and Livestock in the Development Ministry.

The paucity of published material on or by the Central Agronomy Station indicates a low level of activity until the Agricultural Experiment Station at La Molina was created in 1924. The La Molina experiment station received earmarked funding from a 5 percent additional tax on alcohol and alcoholic beverages, plus a voluntary five-centavos tax on each sack of ***guano de isla*** (bird manure) fertilizer. The National Agriculture Society was instrumental in founding the station and directed its activities during the first six years. Legislation converted the various single-crop stations to substations of the central station at La Molina. Later publications indicate this legislation was not fully implemented.

The first scientists with doctoral degrees were Charles H. T. Townsend, an entomologist who began work at La Molina in February 1926, and Ernesto V. Abbott, a phytopathologist who started the following January. The station administration was formally organized in August 1927 with Ing. Gerardo Klinge A. as superintendent.

Meanwhile, the National School of Agriculture at La Molina had been graduating students since 1905. From the beginning, the school administration indicated a desire to do practical agricultural research but was usually without the resources to accomplish very much. However, very early on, Ing. Jorge Vanderghem, the director-founder of the school, diagnosed the potato disease *hielo* (Phythophthora infestans), determined its cause, and prescribed a control. He received a Municipal Gold Medal in recognition of this achievement. From 1933 to 1941, school and station were brought closer together when both were administered under the Peruvian Institute of High Agricultural Studies. Since 1941, each has been a completely separate entity.

From the very beginning, experiment station publications covered a wide variety of crops. Examples of early potato work include a 1928 publication describing the symptoms of two common potato diseases and prescribing controls. A 1929 publication describes common insect pests and prescribes controls. Another in 1931 describes disease symptoms of several cultivated plants and prescribes control measures. In 1936 the station reported on germplasm collection and classification work in the southern Sierra of Peru. A 1939 publication discussed various aspects of potato production and, in particular, makes recommendations on the application of chemical fertilizer. A 1944 publication stands as a landmark in Peruvian potato research. It summarized and synthesized the results of some twenty field experiments conducted between 1928 and 1943. A second 1944 publication outlined a genetic plan to produce potato seed. Most of this work was conducted by Peruvian scientists, though the names suggest that one or two authors were expatriates.[3]

SCIPA and SIPA from 1943 to 1968

Formal agricultural extension work got underway in Peru in 1943 with the founding of the Inter-American Cooperative Food Production Service (*Servicio Cooperativo Interamericano de Producción Agropecuaria,* or *SCIPA*). SCIPA was typical of the services founded across the continent by joint U.S.–Latin American efforts.

> They emphasized production of local food crops, and the creation of rural agricultural services which would institutionalize the technical transformation of the Latin American economy. U.S. self-interest was written [all] over this . . . since the Second World War was on, and secure sources of basic foods for rubber workers in the jungle, for urban centers, and for U.S. garrisons, were considered essential to U.S. interests.[4]

By 1956 a full-blown program of research was under way in the country. In his study of the problems of potato cultivation in Peru, U.S. adviser Fred Cochran listed the experiments in progress. Many were being uniformly replicated at several locations throughout the country (Table 2.1).[5] Cochran concluded his study with a series of observations and recommendations. The following year most of his suggestions were incorporated into the potato section of the National Research Plan developed by the Cooperative Agricultural Research Program. The plan also included rice, corn, beans, wheat, pastures, and livestock. Sugarcane and cotton were excluded because it was felt that existing research efforts were sufficient. The recommended 1958 Peruvian budget had much greater increases for potato research than for any other crop. Potatoes were clearly receiving priority attention in research.

There was also some altruistic spirit, especially after the war when the programs were designed to develop the national agricultural economy. With the passage of time, the U.S. missions gradually shifted their emphasis from an operational to an advisory capacity. At the end of 1960, an expanded research and extension agency, Agricultural Research and Extension Service (*Servicio de Investigación y Promoción Agropecuaria,* or *SIPA*), was formed. SIPA was a semiautonomous agency under the Ministry of Agriculture with four technical divisions: Agricultural Research, Agricutural Extension,

Table 2.1. Potato Research Under Way in Peru in 1956

1. Systemic insecticides
2. Comparison of insecticides
3. Time of application of insecticides
4. Influence of climate and soil on the "Anguilula Dorado"
5. Varieties and hybrids of Puno
6. Acclimation of foreign varieties
7. Material for crosses
8. Creation of varieties for hybridization
9. Collection of varieties and species
10. Comparisons of yields
11. Study of resistance to *hielo* (P. Infestans) of the potato
12. Maturity of seed
13. Weight of seed and density of planting
14. Study of "Guano de las Islas" supplemented with phosphorous and potash in the cultivation of potatoes
15. Preservation of tubers
16. Production of seed of selected potatoes
17. Study of virus in the cultivation of potatoes
18. Comparison of fungicides
19. Study of virus in seed

Source: Fred D. Cochran, *Estudio de los Problemas de Cultivo de la Papa en el Perú* (Programa Cooperativa de Experimentación Agropecuaria, Informe Especial No. 9, Lima, November 1956).

Agriculture Promotion, and Livestock Promotion. In 1968 SIPA was taken more directly into the Ministry of Agriculture where it became the Agricultural Research Directorate and the Agricultural Extension Directorate. These directorates functioned until the ministry was divided in late 1974 into the Ministry of Agriculture and the Ministry of Alimentation (food).

The extension program grew slowly but steadily until it was completely revamped in 1974. It is estimated that by 1970, 151 agency and 522 sector offices were providing services to 102,500 (about 15 percent) of the 850,000 agricultural units in the country.[6] The programs were patterned on the U.S. extension model—there were home improvement programs, rural youth programs, and the core agricultural programs. While extension received somewhat less attention than did research, it was always a component in the North Carolina Mission. At times there were several full-time extension advisers. In later years, supervised credit became an important component of the total extension program.

Technical Assistance by Land-Grant Universities

U.S. land-grant universities were a significant part of the assistance to Peruvian agriculture. During its active years, the North Carolina State

University (NCSU) mission provided the most concentrated effort in technical assistance. It began with the signing of the first contract in November 1954. By late 1955 seven technical experts were on the job in Peru; the number remained at about that level until 1963 when technical personnel increased substantially to some twenty-six.[7] The staff level remained at about twenty-six to thirty scientists until 1969, when the number was reduced to twenty-two. Thirteen people were on board in 1970; the number had declined to seven in 1972 when the mission was closed.

Most of the North Carolina personnel operated as advisers and were assigned to Peruvian counterparts. They helped design and conduct experiments that were integral to the projects of the Ministry of Agriculture. An agronomy adviser, for example, might conduct field experiments, do laboratory analyses, do statistical analyses of the results, and then prepare a report on the work—elbow-to-elbow with his or her counterpart. This ideal was not always possible, as for example, when a counterpart was studying out of the country. In some cases, the advisers would take responsibility for teaching a course, or less frequently, assume administrative responsibility for a program.

Scholarships for M.S. and Ph.D. degree programs and specialized shorter training programs were administered by the mission. By the end of the contract, almost every counterpart, major administrator, and many other technical staffers had received training of some kind, often at North Carolina State in Raleigh, or in another land-grant university, or in the USDA. In the process, they were also exposed to a wide cross section of producers, processors, and marketing institutions. Thus a good number of Peruvian scientists became very familiar with modern U.S. agriculture.

There was also support to construct and furnish buildings, laboratories, and other facilities. Peru's National Agrarian University (UNA) campus was significantly expanded by a building program supervised by NCSU. A great deal of modern laboratory equipment was purchased and installed through efforts of mission personnel. While the Peruvian institutions operated on their own budgets, the technical assistance contracts were instrumental in extending many programs, sometimes by setting an example and sometimes by providing the critical extra resources.

Over the eighteen-year life of the North Carolina Mission, some 270 scientist-years of full-time U.S. technical assistance were provided to Peru. As many as ten or a dozen experts visited the country on short-term assignments each year for at least ten years. In addition, during the last half dozen years, there were six to eight campus coordinators in Raleigh with responsibility in various technical fields. The number of

Peruvians receiving advanced training (M.S. and Ph.D.) under mission sponsorship certainly exceeded a hundred and may have reached two hundred or more. Mission budgets were estimated to have ranged from $250,000 to $2 million annually for technical personnel and their support.[8]

Mission efforts were initially focused on the research and on extension agencies of the Ministry of Agriculture. In later years, substantial assistance was also extended to UNA at La Molina. North Carolina advisers apparently never felt that the separation of the teaching function at UNA from the research and extension function in the ministry constituted a serious obstacle. Or perhaps they saw no realistic way of merging all three under one administration. In any event, the principal research station and UNA are located just across the street from one another in La Molina. University faculty were always somewhat aware of what their research colleagues were doing.

The advisers and the Peruvian trainees represented all the disciplines commonly found in a U.S. college of agriculture together with home economics. Mission personnel worked primarily with nonindustrialized crops and livestock. Accordingly, the appropriate biological sciences and agricultural economics received greater emphasis than other disciplines. The mission constituted a sustained and extensive effort by North Carolina State University.

Iowa State University had an assistance mission in Peru from 1962 to the late 1970s. It was smaller and more specialized than the North Carolina effort, dealing primarily with legal and economic aspects of agrarian reform. During the 1960s, in particular, Iowa State economists conducted numerous feasibility studies on the potential reform targets. During the 1970s, there were evaluation studies of the reforms plus studies covering most of the economic facets of Peruvian agriculture. The Iowa mission was a substantial source of additional technical assistance.

Potato Program from Inception to 1976

The early research at the experiment station at La Molina was organized according to academic disciplines rather than by commodities. Each unit worked on a variety of crops. There does not seem to have been a single station or unit devoted to potatoes, though there were units devoted to several of the agroindustrial crops, such as sugar, cotton, and grapes. In 1957 or 1958, probably with the inception of the National Research Plan, several commodity programs, including a potato program, were established in the Ministry of Agriculture. Disciplinary research continued but was focused, coordinated, and augmented by the commodity programs. The programs survived various reorganizations of the ministry and continued into the 1980s. The potato research and promotion program expanded gradually but steadily during its first decade. Table 2.2 lists the potato research reported in the 1967 *Memoria Anual* of the La Molina experiment station. The listing, which covers most aspects of potato improvement, reveals a broad effort. Some eighty pages were required to report the results of the 1967 work. As with earlier years, many of these experiments were uniformly replicated in the important producing regions of the country. Table 2.3 outlines the programmed research for 1975.

Research results are archived in the library of the experiment station at La Molina but have only been reported in a haphazard output of technical bulletins and extension publications. The results of the hundreds of experiments that have been conducted since 1944 were

Table 2.2 Potato Research Under Way in Peru in 1967

1. Germplasm—collection and introduction
2. Maintenance of germplasm
3. Evaluation of germplasm
4. Development—new varieties—selection of hybrids
5. Propagation of hybrids
6. Comparison of yields—hybrids and varieties
7. Effect of the number of sprouts on production and the number of sprouts on the hybrid Renacimiento
8. Chemical control—effect of the class of disinfectant used in cut potato seed
9. Foundation seed producers—nucleus of foundation
10. Physiology—effect of removal of flowers on hybrid Renacimiento
11. Effect of weight of cut seed on production and number of tubers
12. Comparative study of preemergent herbicides in potato cultivation
13. Number and time of hillings
14. Effect of height and time of hillings
15. Physiology—effect of weight of cut seed on production and number of tubers in the Renacimiento hybrid
16. Type of cut of tubers of the Ticahuasi variety
17. Use of preemergent herbicides
18. Comparison of fungicides used on cut seed
19. Effect of the type of eyes left in the seed tuber on production and number of tubers
20. Comparative study of foreign and national varieties
21. Foundation seed producers and projects applied to foundation seed
22. Chemical control—use of sprouting accelerators
23. Tuber index
24. Cultural practices—number and time of hilling by seed depth

Source: Memoria Anual (Estación Experimental Agricola de la Molina, La Molina, Lima, 1967). This is an edited index of the potato section.

Table 2.3. Potato Research Programmed for 1975

Improvement:
1. Evaluation of hybrid selections
2. Comparison of hybrids and varieties for yield and other characteristics

Physiology:
1. Chemical control of weeds in potato production
2. Use of growth regulators and their effect on yield in potato production
3. Storage of consumer potatoes treated with sprouting inhibitors
4. Storage of seed potatoes in central coast conditions

Seed Potatoes:
1. Obtaining mother plants of basic seed
2. Maintaining and multiplying mother plants
3. Symptomatological determination on the coast of virus diseases in basic seed potatoes
4. Production of seed on the coast
5. Spread of virus diseases on the coast
6. Study of potato degeneration due to environmental factors

Agronomy:
1. Study of response to NPK
2. Agronomic factors in the yield of early and medium varieties (sources and times of application of N and P). Time of hilling of optimal densities for total yield and large-size tubers
3. Comparison of new hybrids and varieties in new irrigation conditions
4. Agronomic practices on production from native seed

Bromatology:
1. Nutritive quality of improved cultivars for the central coast
2. Culinary quality of the varieties grown on the central coast

Ecology:
1. Planting times for different varieties of potatoes

Entomology:
1. Systematic evaluation of harmful potato insects
2. Systematic evaluation of biological controls of potato diseases
3. Determination of forms of selectivity in the control with pesticides of the principal potato diseases

Phytopathology:
1. Evaluation of commercial varieties and advanced selection for resistance to fungal *hielo*
2. Incidence of fungal *hielo*, time of planting
3. Effect of crop rotation on the level of incidence of *Thecaphora solani*
4. Distribution and incidence of *Thecaphora solani*
5. Evaluation of commercial varieties and germplasm for resistance to *Thecaphora solani*
6. Observations on the behavior of seed from the coast vs. seed from the sierra with respect to the level of incidence of Marchitez fungus
7. Storage diseases

Nematology:
1. Response of some potato varieties to chemical control of rootnode nematode *Meloidogyne sp.*
2. Selection of native varieties of hybrids in search of resistance to rootnode nematode
3. Effect of rotations of potatoes with plants auto-genic and inhospitable to the development of populations of the rootnode nematode
4. Integrated teamwork in the nemotological evaluations of the project

Source: Proyecto de Papa (Ministerio de Alimentación, Dirección General de Investigación Agraria, Centro Regional de Investigación Agraria, Estación Experimental Agraria, La Molina, Lima, July 1975).

not, in the late 1970s, organized, summarized, or synthesized into publications suitable for either a technical or a producer audience.

Excerpts from the *North Carolina Mission Reports* characterize the spirit and direction of the "In-depth Potato Project" from 1967 to 1970. The new program identified three key areas for potato improvement: adaptive research, seed production, and extension. The mission supplied the project with a co-leader, an extension adviser, a seed production adviser, and also backstopped the project with a mission plant pathologist, as well as campus coordinators in soils, plant breeding, and physiology.

> If potato production is to be increased, storage facilities developed, and adequate marketing facilities established in Peru, the problems affecting these operations must be solved through adaptive research.[9]
>
> The long-term goal is to leave behind an institutional capacity to continue the development of all phases of the program and hence to have an important impact on the production of this very, very key food commodity in Peru. It has been estimated that four or five years will be required to have a marked impact on the commercial seed producers. . . . Highlighting the six-month period (first half of 1970) was the participation by twenty Peruvian scientists at the meetings of the Latin American Society for Potato Research. The quality and quantity of the Peruvian work impressed everyone attending.[10]

Other Influences on Potato Research

International Potato Center

At the time the North Carolina mission was winding down, the *Centro Internacional de la Papa* (CIP) was being established at the same location. Several members of the mission staff who were involved with the potato program were instrumental in founding CIP. The first director general, one member of the governing board and several senior scientists all moved from NCSU over to CIP. It should be noted that the aims of the two institutions were very different and that CIP was in no way a replacement for the North Carolina Mission. CIP is a nonprofit, autonomous scientific institution established by agreement between the Consultative Group on International Agricultural Research (CGIAR), and the Government of Peru to promote greater use of the potato as a basic food.

As one of the worldwide network of centers funded by CGIAR, CIP has the following mandate:

> To develop, adapt, and expand the research necessary for the technology to solve priority problems that limit potato production in developing countries. This includes adapting the collective knowledge that has contributed to the stepwise increase in potato production in developed countries.
>
> The world potato collection, which has been accumulated by CIP, provides ample opportunity for research breakthroughs through the exploitation of this large quantity of previously unavailable genetic resources.[11]

In Peru, CIP's work is conducted at four main experimental sites: the irrigated desert at La Molina, the highlands at Huancayo, the mid-elevation tropics at San Ramón, and the low humid tropics at Yurimaquas. These sites represent all the major production zones in the country.

CIP gathers and maintains an extensive germplasm collection, provides genetic material to national breeding programs, does extensive research on potato diseases and pests, and trains potato scientists from around the world. Located immediately adjacent to the agricultural experiment station at La Molina and directly across the road from the National Agrarian University, its laboratories and experts are directly accessible to Peruvian scientists. Potato technology tested under Peruvian conditions has now reached a world-class level.

Agrarian Reform

Another event occurred in Peru from 1969 through the 1970s that had a profound impact on Peruvian agriculture. The national agrarian reform program, one of the most thoroughgoing in Latin America, is difficult to assess in any context. In brief, the reform expropriated the large haciendas of absentee owners and converted them into production cooperatives. The cooperatives were not solidly institutionalized before a new and less sympathetic government replaced the revolutionary military regime. There is evidence indicating that while the reforms radically transformed land ownership patterns, long-term production trends were altered very little—at least until the late 1970s. Potato and other agricultural production increased up until 1980 at about the same rate (2 percent) that preceded the reform. Since 1980, severe droughts and floods have sharply reduced production in several commodities. In sum, the reforms probably had very little impact on the rate of adoption of agricultural technology, at least during the early 1970s. So much of Peruvian agriculture had been so indifferently managed previous to reform that even its disruptions did not appreciably reduce production.[12]

The Peruvian Institutions

Notwithstanding complaints of mission advisers, a review of ministry and university documents indicates that the day-to-day Peruvian institutions accepted most of the advice they received. The mission provided an architectural staff to help design and supervise construction of a new section of the UNA campus at La Molina. A substantial social science program was established, and the entire curriculum was broadened and North Americanized. The level of training for the faculty was enhanced when many went abroad for advanced degrees. La Molina graduates performed very well in their professions, and by the 1960s and early 1970s UNA was generally recognized as one of the very best agricultural schools in Latin America. By any standard, UNA improved considerably during the North Carolina presence.

As a postscript, about the time of the mission departure in the early 1970s, a series of political and physical events shook the La Molina campus. The normally conservative and tranquil institution was closed for a period of time by unrest and strikes led by the recently expanded social science component. The government responded by reducing the social science program. A ceiling on all faculty salaries, plus higher salaries elsewhere, led to the departure of many of the more highly trained faculty. The reduced support for UNA coincided with the decline in support for agricultural research institutions in other Latin American countries. Trigo, Pineiro, and Ardila have documented the deterioration of these vital national links to the CGIAR institutions.[13] However, the causes may have been somewhat different in Peru, which was governed at that time by an unusual military regime. The regime apparently felt UNA was too closely tied to the former agrarian structure to contribute to future changes. A major earthquake in October 1974 punctuated these political storms by severely damaging much of the campus. A number of buildings, including several designed and built by the North Carolina Mission, were no longer fit for use. By the late 1970s, UNA was still respectable, but it had clearly declined from the institution it once was.

Measured by research output, the institution-building effort within the ministry was also successful. Peru's research effort grew steadily from very modest beginnings. Both capacity and output increased markedly during the North Carolina presence.[14] Research programs expanded in the directions suggested by advisers, commodity programs were established to concentrate attention on key crops, new facilities were constructed, and many Peruvians received advanced training abroad.[15] A review of experiment station reports shows rather impressive research results. Hundreds of experiments were conducted on most aspects of crop and livestock production. The well-balanced research dealt with germplasm, nutrition, sanitation, pest control, management practices, and other aspects of agricultural production. While much of the research has not been systematically summarized or reported, a great deal of agricultural technology has been developed or adapted to Peruvian conditions. As was mentioned above, extension capacity and effort were also increased markedly during the same period to a level commensurate with the research organization.

For several years in the 1970s, the Ministry of Agriculture was actually divided into two ministries: the Ministry of Alimentation for promoting and assisting food production, and the Ministry of Agriculture with primary responsibility for agrarian reform and assisting the newly established production cooperatives. At the beginning of the 1980s, the two ministries were recombined. The almost continuous reorganization and shifting political alliances led to the departure of many highly trained people. As with UNA, the ministry's capacity for effective work was affected by such attrition. However, it is still legitimate to examine its institution's effect on production at least into the early 1970s.

Some of the technology has reached a few producers. Some are using new varieties, fertilizer, and pesticides. Yet the effect of the new technology is still very limited.

Lagging Agricultural Production

A careful review of Peruvian agricultural production data shows, with few exceptions, a lackluster performance before, during, and since the presence of the North Carolina Mission. Commodity yields most emphasized by the mission were low by world standards, leaving room for substantial improvement. However, neither national yields nor production has increased markedly during or since the mission presence. For example, potatoes received major attention from both donor and recipient, but national average potato yields increased only 6 percent from 1954 to 1972. In a somewhat later period, yields increased by approximately 14 percent (Table 2.4). Yields actually fell by some 25 percent from 1954 to 1972 in the biggest potato-producing regions (representing approximately 48 percent of the national area).[16] These regions were not ignored. In fact, research was conducted in them for many years and they have usually received much attention.

Because of the efforts of all the scientists in the U.S. missions, the UNA, the ministry, and CIP, potato technology, itself, has undeniably advanced. Peru has had hybrid potatoes for almost thirty years.[17] Advances have been made toward controlling some of the more important diseases and pests. Proper fertilizer applications have been determined, and other cultural practices have been tested. With the new technology, a few commercial Peruvian growers have obtained yields that are ten times the national average of six tons per hectare. Over the years, however, potato production has kept up with increased national consumption largely through increases in the acreage under cultivation.

The potato situation is not unique in Peruvian agriculture. Yields of most other major commodities have increased at similarly low rates, while acreage planted and livestock have expanded slowly, if at all. There were three significant exceptions to the general trend. Sugarcane yields exceeded those in the United States, while rice and cotton yields matched U.S. figures. These commodities have long been produced in good measure for export, such that demand and prices were established in the world market rather than in the local market. As was pointed out earlier, research in cotton and rice began in 1901 and in sugarcane in 1906—all at the producer's initiative. Thus, the users of the technology had been interested in it even before it was developed. Improvement of the subsistence commodities, on the other hand, was initiated to a much greater extent from the top and even from outside Peru. There are several reasons why producers have been slow to adopt the new technology.

Overall, national agricultural production has lagged far behind the population growth rate (Table 2.4). Peru's per capita production performance is the lowest of the major countries on the continent. Even during the years of greatest agrarian reform activity (1969-1977), national agricultural production has continued along established trends with virtually no interruption. The Peruvian revolution, which began in 1968, was clearly not a Green Revolution. Overall, economic growth averaged between 5 and 6 percent, while agricultural growth was barely 2 percent, and population was growing at about 3 percent annually.[18] Since agricultural production has not kept pace with the population increase, food imports have increased steadily.

Table 2.4 Peru: Indices of Total Food Production, Per Capita Food Production, Population, and Average Potato Yields, 1961-1979

Year	Index[a]			
	Total Food Production (%)	Per Capita Food Production (%)	Population[b] (%)	Potato Yields (T/ha)
1961-65 Average	100	100	100	5.8
1970	116	94	122	6.1
1971	121	96	126	6.1
1972	113	87	130	6.5
1973	112	84	134	6.8
1974	115	83	138	6.9
1975	113	80	142	6.7
1976	117	80	146	6.4
1977	116	77	151	6.3
1978	106	68	155	6.5
1979	103	65	159	7.0

Source: Production Indices, Agricultural Attaché, U.S. Embassy, Lima. Potato yields, *Production Yearbooks* (Food and Agriculture Organization of the United Nations, Rome).

[a] The indices are each calculated on the base period 1961-1965.

[b] 1961-65 population = 10,902,000.

A few more pieces of information may help interpret these findings. Peru has been self-sufficient in potatoes for many years, importing and exporting only small quantities from year to year. Potatoes are the principal starch, although bread and rice are also staples. Nevertheless, Peruvians eat a lot of potatoes: their per capita consumption is about 100 kg per annum. In fact, table consumption seems to change very little over time or with changes in income. The scarcity of potatoes in Lima during some summers seems to be a temporary seasonal phenomenon. On an annual basis, Peru appears to have an equilibrium market situation with producers offering about what consumers want to buy at current prices over the country as a whole. Very few potatoes are processed.

Conclusions and Implications

The North Carolina Mission and the cumulative U.S. effort in general must be given high marks for institution building. By 1972, the National Agrarian University at La Molina had assembled a well-trained faculty with very good classrooms, laboratories, and offices. The students performed well in their work and in graduate schools outside the country. La Molina enjoyed an excellent reputation throughout

Latin America.

Research and extension capability in the Ministry of Agriculture also reached a relatively high level. Potato research illustrates what was taking place in a number of commodities. In most important disciplines, there was at least one doctoral-level staffer supported by other well-trained people. Laboratories were equipped, and there was a national network of experiment stations that produced locally adapted technology.

Both the university and ministry accepted most of the advice offered by their U.S. advisers, although it is true they have never united in conformance with the American land-grant model. If there was any serious attempt to persuade or pressure the Peruvians into such a merger, it is not evidenced in mission reports or other documents. Few, if any, advisers seemed to regard the formal division of responsibility as detrimental to institutional effectiveness.

It is important to note that benefits from the technical assistance work have accrued to the donor, as well. For example, a New York State potato breeder who has consulted at CIP since its early days used Peruvian germplasm in his domestic breeding program for a number of years. This germplasm provided important sources of resistance to a number of maladies faced by U.S. growers in various parts of the country. For example, the variety *Rosa*, which was released by the New York Experiment Station in 1981, incorporated Peruvian germplasm. This germplasm appears in about 25 percent of the clones under evaluation at that station.[19]

Peru is in the very enviable position of having a substantial body of research—locally adapted technology—ready for use in potato fields. This technology constitutes a valuable but largely underused national resource. The potato yield in Peru has been rather checkered over the past quarter century, increasing slowly in some departments and declining in others. While yield is certainly not the only measure of improved technology, it is one of the best indicators of its benefits. Yield data are also readily available.

Constant yields do not indicate that new technology was absent. They do suggest that few people used it *or* that the effect was masked or offset by other factors. It is quite possible that the most conspicuous producers—those close to Lima or along the roads elsewhere—were the users of the improved technology. Their location may create the illusion that use of better technology is more widespread than it actually is. Yield declines in some areas offset the modest gains made in other regions of the country. It is quite likely that the same thing occurred within regions. The most persuasive explanation for the yield declines appears to be that the expanding population of subsistence producers is being forced into increasingly marginal soil and climatic areas. Also, the fallow period in many areas has been reduced or even eliminated. This also decreases yields unless fertilizer is applied to compensate for the depleted nutrients.

Peru has data from more than fifty years of research representing hundreds of experiments replicated throughout the major producing regions. The research effort needs to continue, since many problems are never really solved—only ameliorated in some degree—and new problems are continually popping up. In view of the low yields in even the best potato-growing area, Ica, it does seem clear that Peru has not made full use of the existing technical know-how. This could be, in part, due to the slow and incomplete reporting of research results as discussed above. However, information from sugar, cotton, and rice research was disseminated without a significantly different reporting system. Something more seems to be required to increase production or improve production efficiency.

In an address before the 1976 World Food Conference at Iowa State University, Clifton Wharton made an observation that is surely apropos of the North Carolina Mission in Peru:

> There is a very sizable consensus within the various professions and disciplines as to what technically is required to achieve significant sustained agricultural development. [However], few nations or regions have achieved the levels of agricultural development which match our state of knowledge in the "art" . . . we have not come even reasonably close to what we know to be possible.[20]

Why is it that the enhancement of institutions and the improvement of technology have not translated into increased agricultural production? Thirty years have passed since the North Carolina effort began in Peru. Three decades exceed the twenty-five-year figure estimated as necessary for institutionalization as discussed in the Floras' chapter. Ten years of U.S. technical assistance preceded the North Carolina effort, and the national institutions had been established forty years before that. What has been missing? In addition to the biological technology, what else is required to increase agricultural production?

An elementary analysis indicates that increased agricultural production requires inputs such as seed, fertilizer, and pesticides. It requires markets. It requires incentives or prices that are sufficiently attractive to motivate the producer. It requires two-way transportation, since production and consumption points are usually far apart.[21] Thus, technology, inputs, a market, producer incentives, and transportation are all essentials for increased agricultural production. If even one of these essentials is absent, no increase will occur. In addition, credit and extension education, while not absolutely essential, do facilitate increased production.

The technical assistance missions focused heavily on biological technology and on credit and extension education, all of which are relatively inoffensive to most interest groups in the host country. Decisions on the use of the new technology are usually made at the farm level and rarely occasion national debate or crisis.

The other essential elements in the agricultural production equation are somewhat different from technology in that they often require national policy decisions. And these decisions often occasion national debate and crisis. For example, agricultural price policy, which establishes producer incentives, is frequently an extremely sensitive issue. High food prices may motivate farmers to produce, but they often motivate urban consumers to riot at the palace gates. Fertilizer or tractor imports may mean balance-of-payment difficulties. Rural road construction may take funds away from urban freeways, and so on. In short, at least a few tough, and sometimes dangerous, political decisions are usually required for agricultural development to occur.

North Carolina and Iowa mission economists, among others, produced a number of studies on marketing, transportation, inputs, and prices, and many Peruvians obtained advanced degrees in these subjects. So, the Peruvian government was not totally unaware of the nonbiological essentials for increasing agricultural production. Consciously or unconsciously, the government elected to go with alternative policies. Among other options, commodity prices were kept low, and adequate fertilizer was not imported. Nor did the technical advisers push very hard on the sensitive political issues. They preferred to remain aloof from the hard political decisions required for successful agricultural programs. Studies were completed and archived, but the findings and policy implications went unheeded. Wharton stated the problem this way: "We in the United States have never really faced up to the political dimension of the agricultural development process in the developing world."[22]

He argues that the political dimension of development cannot be ignored, however sensitive it may be. Peru received and supported biologically oriented technical assistance. Yet it did not make much progress with the other essential elements of agricultural development. The nation pursued a cheap-food policy which meant low incentives for its producers; and agricultural production has lagged as a result.

Peru provides a classic example for the issues discussed in the preceding chapter.

High Stakes

The foregoing should not be interpreted as an argument for direct political involvement by technical advisers or for less sovereignty for host countries. It is neither. However, technical assistance missions, research stations, and developing countries can no more afford to ignore essential political elements than can the individual producer. Successful farmers may not actually control prices and the availability of inputs, but they keep a close and continuous watch on each. They line up their supplies in advance of the planting season. They line up transportation, be that a few burros, a neighbor's truck, or several rail cars. They make implicit or explicit assumptions about the market and mentally calculate whether they want to produce at those prices. They may even hedge their produce to ensure a minimal profit. All successful producers follow these practices routinely, and concern themselves, as well, with the technical aspects of production—such as which variety to plant or how much fertilizer to apply. When the market price of a given commodity falls too low or when some other essential element is absent, the farmer stops producing that commodity, shifts to other commodities, or goes out of business.

Technical assistance missions might be well advised to operate in the same fashion. When one or more essentials of the market equation cannot be put into place after a reasonable time, it might be wise to terminate the effort and to spend the money elsewhere. Governments that *can* control the essentials must be willing to do so, or production cannot be expected to increase.

Many technical experts would probably agree that development is complex and that certain political and economic supports are essential. Most actual programs, however, have been designed to deal almost exclusively with biological technology. It was expected that increased production would follow. Current discussions of the Green Revolution exhibit this oversimplification of reality. The high-yielding grain varieties have been touted by many as the answer to the world food problem. Even a casual reading of case studies, however, shows that where production increases, it is not with the introduction of the variety alone, but with *all* of the above elements in place.[23]

Nobel Laureate Norman E. Borlaug has recognized for years that the major obstacles to increasing food production are no longer scientific.

> If it's [increased food production] going to work, the technical package has to be linked to economic and political policy. The people in policymaking positions must have the relevant facts. Generally, however, from their own bureaucracies they never get the facts.[24]

The case of potatoes in Peru is relevant for major technical assistance programs throughout the world. The lesson is clear. Scientific advancements can be made, but if they are not coupled with proper economic and political policies, the benefits will not go beyond the research centers.

Technical experts in both donor and host institutions would do well to remember how the world appears to a producer. Programs to increase agricultural production must, of necessity, take into account all of the elements in the production equation. A whole supportive environment has to be created. Efforts that ignore even one vital element run the risk of failure, just as any farmer who neglects even one aspect of his business may also fail.

We must face the basic question of how much technical assistance to extend in the absence of adequate political and financial commitment. Should donors continue to extend technical assistance to countries that do not develop adequate food policies, expecting few, if any, increases in agricultural production? Or should donors be very hard-nosed and extend assistance only to those willing and able to make the tough decisions and formulate the appropriate policies? There are probably only a few such countries in the world at any one time.

If donors push for control, they infringe on the sovereignty of the host country. But, if they back away from the tough and sensitive issues, they may lull the developing countries into thinking technology alone will provide an easy or magic solution. Without a progressive agricultural environment and without all essential elements in place, agricultural production will not increase. The stakes are high. The reward for achievement is human well-being, and the penalty for failure is widespread malnutrition and famine.

Notes

1. This chapter draws heavily from Eastman (1976, 1977).

2. This entire section is drawn from the early *Memorias Anuales* of the Estación Experimental Agricola, and the *Memorias* of the Escuela Nacional de Agricultura y Veterinaria (La Molina, Lima), as well as from J. Alberto León et al. (1952).

3. See Abbott (1928, 1931); Wille(1929); Soukup (1936); Eckert (1939); Calzado Benza (1944); and Boza Barducci (1944).

4. Rice (1977).

5. Cochran (1956). Similar studies were done on all major agricultural commodities and published as a set of special reports.

6. *La Problemática del Desarrollo de la Agricultura Peruana* (Lima: Ministerio de Agricultura, 1971). Agencies were located in most important towns and cities around the country and were staffed by university-trained personnel. Sector offices were located in villages in important agricultural areas and were staffed by personnel with training in vocational agriculture.

7. In 1955, a chief of party, a soil scientist, and a geneticist were located at La Molina. An adviser to the director, a livestock specialist, an agronomist, and a physical plant specialist were located at Tingo María.

8. Estimate of Milton W. Lau, former chief of Agricultural Division, USAID, Lima, personal communication.

9. *Semiannual Report* (North Carolina State University Mission to Peru, January-June 1967).

10. *Semiannual Report* (North Carolina State University Mission to Peru, January-June 1970).

11. International Potato Center, *Annual Report* (Lima, 1982).

12. See Eastman (1984).

13. See Trigo, Pineiro, and Ardila (1980), and Ardila, Trigo, and Pineiro (1982).

14. Clyde Eastman, "Potato Research and Potato Yields in Peru" (Socioeconomic Paper No. 2, International Potato Center, Lima, June 1976).

15. The commodity programs are described in French and Apple (1974).

16. Regional production data are from Estadísticas Agrarias (prepared annually by the Ministerio de Agricultura, Lima), and from the U.S. Agricultural Attaché in Lima.

17. High-yielding, disease-resistant potato varieties are developed by crossing different germplasms (hybridization) to obtain the desired combination of characteristics. Once the desired qualities are obtained by sexual reproduction, the variety is thereafter multiplied vegetatively.

18. These data come from *Cuentas Nacionales del Perú* (Banco Central del Perú, Lima) and from the American Agricultural Attaché in Lima.

19. Telephone interview with Dr. Robert Plaisted, Cornell

University, March 5, 1984.

20. Wharton (1976).

21. This deceptively simple conception of agricultural development has been outlined by Mosher (1966).

22. Wharton (1976).

23. For example, see Vyas (1975).

24. Wolff (1983).

References

Abbott, E. V. *Scab y Hielo de la Papa.* Estación Experimental Agricola, Circular No. 7, Lima, Marzo 1928.

——. *Enfermedades de las Plantas Cultivadas en el Perú.* Estación Experimental Agricola, Circular No. 18, Lima, Deciembre 1931.

Ardila, Jorge, Eduardo Trigo, and Martin Pineiro. "Human Resources in Agricultural Research: Three Cases in Latin America." *Agricultural Administration* 10 (1982):213-34.

Boza Barducci, Teodoro. *Plan Genético para la Producción de Semilla de la Papa.* Estación Experimental Agricola, Boletín No. 24, Lima, Noviembre 1944.

Calzada Benza, José. *Resultados de la Experimentación Sobre el Cultivo de la Papa.* Estación Experimental Agricola, Circular No. 62, Lima, Junio 1944.

Cochran, Fred D. *Estudio de los Problemas de Cultivo de la Papa en el Perú.* Programa Cooperativa de Experimentación Agropecuaria, Informe Especial No. 9, Lima, Noviembre 1956.

Eastman, Clyde. "Land-grant Colleges and Agricultural Development: Lessons from Peru." Focus section of *International Development Review* 4 (1976):23-25.

——. *Technological Change and Food Production: General Perspectives and the Specific Case of Potatoes.* Lima: International Potato Center, 1977.

——. *Peru's Agrarian Transformation.* Latin American Research Paper Series No. 1, Center for Latin American Studies, New Mexico State University, Las Cruces, April 1984.

Eckert, Gerardo. *Recomendaciones para el Mejoramiento de la Producción de la Papa en la Sierra.* Estación Experimental Agricola, Circular No. 48, Lima, Setiembre 1939.

French, E. R., and J. L. Apple. "The Commodity In-depth Projects of North Carolina State University in Peru: A Model Bilateral Aid Programme." *FAO Plant Protection Bulletin* 22 (April 1974):2.

International Potato Center. *Annual Report.* Lima, 1982.

León, J. Alberto. *La Escuela Nacional de Agricultura en sus Bodas de Oro 1902-1952.* Lima:(UNA), 1952.

Mosher, Arthur T. *Getting Agriculture Moving.* New York: Praeger, 1966.

North Carolina State University Mission to Peru. Semiannual Reports. January-June 1967 and January-June 1970.

Perú, Banco Central. *Cuentas Nacionales del Perú* (various years).

——. Escuela Nacional de Agricultura y Veterinaria. *Memorias* (various years).

——. Estación Experimental Agricola. *Memorias Anuales* (various years).

——. Ministerio de Agricultura. *Estadísticas Agrarias* (various years).

——. Ministerio de Agricultura. *La Problemática del Desarrollo de la Agricultura Peruana*. Lima, 1971.

Rice, E. B. "Extension in the Andes." Washington, D.C.: AID Evaluation Paper 3A. Agency for International Development, April 1977.

Soukup, R. P. Valeriano. *Clasificación de la Papa en el Departamento de Puno*. Estación Experimental Agricola, Boletín No. 18, Augusto 1936.

Trigo, Eduardo, Martin Pineiro, and Jorge Ardila "Aspectos Institucionales de las Investigaciones Agropecuaria en América Latina: Problemas y Perspectives," *Desarrollo Rural en las Américas* 12, no. 1 (1980):3-25.

Vyas, V. S. *India's High Yielding Varieties Programme in Wheat, 1966-67 to 1971-72*. Mexico City: Centro International de Mejoramiento de Maiz y Trigo, 1975.

Wharton, Clifton, Jr. "The Role of the Professional in Feeding Mankind: The Political Dimension." Keynote address presented at the World Food Conference, Iowa State University, Ames, Iowa, June 28, 1976.

Wille, Johannes. *Injurious Insects Which Attack the Potato in Peru* Estación Experimental Agricola, Circular No. 16, Octubre 1929.

Wolff, Anthony, "Borlaug Remembers," *RF Illustrated.* New York: The Rockefeller Foundation, December 1983.

PART 2

SELECTED PROBLEMS

The promotion of agriculture and the improvement of the quality of life of rural populations entails grappling with a large number of challenges. In Part 2 a few of the more important challenges are addressed: the communication of scientific knowledge, women's roles in agricultural development, the integration of the work of research scientists and extension educators, the adaptive transfer of program models, and the linking of scientific and indigenous agricultural systems. This selection of problems and related issues deals largely with the nontechnical side of agriculture which has been relatively neglected.

Colle assesses the conventional systems for relaying information to farmers and discusses the role of extension workers and scientists as communicators. He addresses the principles of message saturation and simplified media and suggests the need for communication support units, better planning, and an aggressive agricultural information network. In essence, he points to areas where much improvement is needed.

Sachs and Caye describe the role of women in agriculture, focusing on their productive activities in relation to both food and cash crops, the effects of new technology on women's status, women's access to land, and the out-migration of males. The authors then discuss factors affecting "women in development" programs, and conclude that policy changes and institutional initiatives have not resulted in substantial increased participation of women. They address some of the factors that account for this lack of progress.

Compton presents a brief history of efforts to develop extension programs in Third World countries, ranging from the diffusion of innovation approach, package programs, induced innovation strategies, to more recent decentralized user-oriented models. He then discusses some of the lessons learned in efforts to effectively link the functions of knowledge creation, diffusion, and utilization in agriculture.

Rogers describes the historical evolution of the U.S. extension model and discusses the efforts made to transfer it to developing countries. He points out the importance of local control, a needs orientation, the role of the subject matter specialist, and the use of indigenous aides within the U.S. system and the comparative absence of such dimensions in the developing countries' extension programs.

Warren assesses scientists' previous lack of interest in indigenous agricultural knowledge and practices and the emergence of such interest during the past two decades. He suggests the implications this is likely to have for efforts of agricultural research scientists and extension educators to increase the agricultural productivity, income generation, and general quality of life of small-scale producers.

CHAPTER 3

Communicating Scientific Knowledge

R. D. COLLE

This chapter examines what is involved in transferring scientific information to farmers and explores how this can be done in a timely way that allows farmers to gain maximum benefit from their agricultural enterprise.

On the surface, this communication process may seem like a moderately simple question of providing an adequate number of extension agents or "packaging" scientific information so that farmers can understand what scientists have to say. However, a review of the strategies used during the past few decades, with a look at the challenges of the communication problems today, and an assessment of the communication resources for tomorrow make it clear that linking scientists and farmers involves more than the packaging and "conventional" extension systems.

The issue is vital because agricultural research can result in substantial increases in quality and quantity of production if farmers adopt the appropriate technologies. For example, in India, scientists have developed technologies that could multiply the yields of pulses and other crops threefold under field conditions, but effective communication is still necessary to achieve those results widely (Prasad 1985).

The task is complex. In a sharply worded indictment of behavioral scientists playing the role of "facilitators," Saint and Coward (1977) remind us of the need to consider a variety of variables and issues in the rapid adoption of innovative practices, including technological and institutional factors and ecological systems. To these we might add questions related to research goals, indigenous knowledge, equity, intraorganizational networks, and linguistics. Several of these topics are addressed elsewhere in this book (see especially the chapters by

Compton and Rogers), so they will be mentioned here only briefly to provide context for our discussion on how to improve the flow of information *between* institutionally-based scientists and the farmers who are potential users of research information.

To begin, it may be useful to go to the field where one kind of research consumer works.

Shahhat and a Changing Agriculture

Richard Critchfield's (1978) story of Shahhat illustrates some of the changes in agriculture that influence our discussion of communicating scientific knowledge. Shahhat, whose name in Arabic means peasant or beggar, lives near the Nile River, where this story really begins. Much of life and agriculture in upper Egypt (south of Cairo) had been controlled for centuries by the river's seasonal flows. In August, the Nile would rise above its banks, only to be restrained by dams built by the fellaheen (Egyptian peasant farmers). The river was allowed to flow through the dams between September and November to deposit a fresh, fertile covering of silt on the nearby lands. The fellaheen would then sow their crops and wait for harvest four months later. Only one major planting was done each year; summer was a time for rest.

Then population pressures led to the building of the Aswan Dam. The dam would control the flow of the Nile so that water would be available year-round, allowing more intensive cultivation. Agricultural production could be increased to meet the food needs of the large and still growing numbers of people. The new goal was to have two or three plantings in succession each year on the same amount of land, thus resulting in higher yields per acre for the year. But these were *national* goals and not necessarily Shahhat's. Critchfield recounts:

> Shahhat was grieved to see the Nile's last flood. From the first, he distrusted Nitrokima, the new chemical fertilizer artificially manufactured at Aswan to replace the Nile's lost mineral-rich silt. Until now only land susceptible to the annual flooding, and the natural re-fertilization it brought could be cultivated. Once the high water came in August and the fields were inundated, there was little to do but sow one's wheat, barley, lentils or maize in November and wait for the April harvest. And now three crops were to be grown each year. Even in the scorching summer heat, the field work was unending, and chemical fertilizer had to be used for the first time in the Nile Valley. . . .
>
> Shahhat felt frustrated. Left to himself to cultivate his father's acre of ancestral land beside his house, he was content. He used the methods handed down from the time of the pharaohs, and everything he put his hand to turned out well. On the government-deeded land at

> Sombat it was different. He had to depend upon Faruk [a fellow sharecropper] or the inspector for almost everything, and things seldom went as promised. He longed to be independent of anyone. (Critchfield 1984)

Shahhat is caught up in a web of change that he finds hard to understand, and even when understood, he finds hard to accept. Shahhat's story highlights some of the changes that in the last several decades have brought new knowledge and technology to the agricultural enterprise, inducting small farmers—sometimes willingly, sometimes unwillingly—into the system. Heretofore these farmers had been overlooked, ignored, or treated condescendingly by governments, educators, scientists, extension officials, bankers, agricultural input suppliers, and the press.

To survive, small farmers like Shahhat must adjust. To adjust, they must know more about the new technology that may be thrust upon them, or merely available to them. Will they seek the knowledge, or will it seek them? Is it accessible? Understandable? Appropriate? Timely? These are a few of the questions central to the issue of how we may speed and maximize the practical payoff of agricultural research.

The task of communicating scientific knowledge about agriculture requires careful review because, as Shahhat's story suggests, important technological and sociological changes are taking place that influence the process. To see these changes and their ramifications, we need to review the pattern evolved over the past two or three decades that still dominates the methods used for moving research information from experiment stations to the farm.

Conventional System for Relaying Information to Farmers

The widely accepted model for transmitting scientific information to farmers usually follows this pattern:

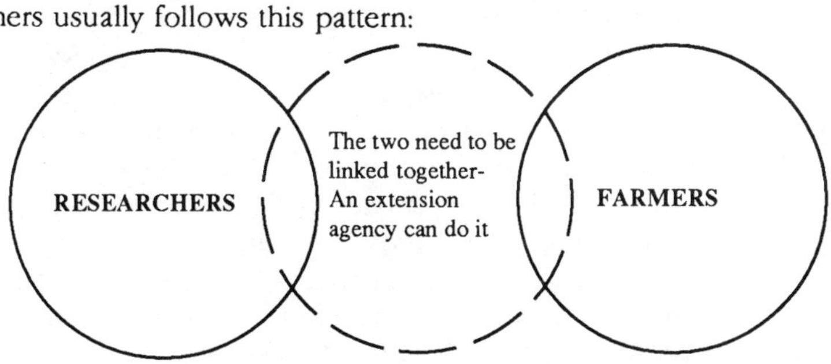

Source: Herbert F. Lionberger and Paul H. Gwin (1982)

U.S. land-grant institutions and social scientists have played a significant role in shaping the communication patterns that have emerged in Third World countries putting into operation the linking structure illustrated above. Awa (1981) notes that after World War II, U.S. extension models spread rapidly to other parts of the world, especially to the newly independent countries of the Third World. Although the forms may be similar, he suggests that these Third World models do not reflect the assumptions of the U.S. system that relate to funding models, the participation of land-grant universities, and the integration of research, extension, and teaching.

Although the organizational relationships for carrying out the functions implied in the diagram above may vary, the basic pattern depends heavily on face-to-face interpersonal contact. Blanckenburg (1984), for example, analyzed extension systems in African and Asian countries using secondary data, and discovered that while a variety of media and methods is used in all extension agencies, the "classic" methods predominate—namely individual advisory and group methods. He cites several studies conducted from the 1960s to the 1980s in which Asian, African, and Near Eastern extension officers have been asked to rank their extension methods in order of importance. Farm and home visits ranked first, followed by meetings, group discussions, and demonstrations. Blanckenburg suggests that extension agents don't use mass media because they (the agents) lack the proper equipment, and because the mass media services are not integrated into the extension service.

In agricultural programs, "an extension agency" has become synonymous with agents fanning out among farmers to promote adoption of research station findings. Extension education and rural sociology studies have generally produced results to confirm and reinforce this fundamental pattern of emphasis on the agent as the main communicator in technology transfer.

The Influence of Diffusion Studies

Perhaps the most influential concept associated with extension and the linking of researchers and farmers is that of diffusion and adoption. A study by Ryan and Gross of the diffusion of hybrid corn laid the foundation for this conceptual approach more than forty years ago (Rogers 1982). Since that time almost 800 rural sociology diffusion studies—mainly related to agricultural ideas—have been recorded at the Diffusion Documents Center at Stanford University. About 20 percent have been conducted by researchers in extension.

Generally these studies indicate that the communication of scientific information involves more than setting up a transmission belt to move information from researchers to farmers or other kinds of consumers. They conclude that the diffusion of innovations involves five stages: knowledge, persuasion, decision, implementation, and confirmation. Similarly, the adoption of innovations may involve five kinds of adopters: innovators, early adopters, early majority, late majority, and laggards. Each diffusion stage has its own communication characteristics (such as the most effective channel), and the adopter categories have distinctive communication and interpersonal network patterns.

A study from India illustrates how diffusion research seems to have been inadvertently responsible for reinforcing certain communication strategies and closing off fruitful alternatives for more efficient flow of information between farmers and researchers. Following a pattern found in many diffusion-adoption studies, 245 farmers were questioned about "the different sources which influenced them at different stages of adoption of the improved processes of hybrid bajra cultivation." The farmers were asked the "utility" of twenty-nine information sources listed in an interview schedule for awareness, trial, and adoption of the new techniques.

Not surprisingly, radio and films had the highest rankings for awareness, with the village level worker's (VLW) office and the agricultural assistant's office ranking third and fourth. The same four sources were the highest ranked for the trial stage, but in reverse order. For the adoption stage, face-to-face contacts, especially those by extension agents (farm and home visits of the agricultural assistant and the VLW) topped the list. A report of the study is worth noting because it draws conclusions on the strategic use of channels for reaching farmers.

> In the present circumstances, large-scale adoption of recommended farm practices is essential to increase the output in farming. To be effective in this process, one must know what techniques and influences can be effectively used during different stages of adoption [sic]. Radio and films are effective in creating awareness whereas adoption *depends on the direct exertions of the Gram Sevak (VLW) and Agricultural Assistant.* (Bajat and Thorat 1978, emphasis added)

One of Rogers' (1982) generalizations about information sources echoes the same idea:

> *Mass media channels are relatively more important than interpersonal channels for earlier adopters than for later adopters.* . . . But the less change-oriented, later adopters require a stronger and more immediate influence, like that from interpersonal networks. [Emphasis supplied]

The scores of diffusion studies and the "generalizations" derived from them reinforce the strategy of employing extension agents as the principal communicants for connecting researchers and farmers. Other channels, such as media technology, are assigned secondary importance. For example, in their recent book *Communication Strategies: A Guide for Agricultural Change Agents,* Lionberger and Gwin (1982) *de-emphasize* the role of media in the extension process by titling a chapter: "Mass Media Channels as Communication Assists."

A major consequence of this historical emphasis on interpersonal contacts is that (1) roles for agents have become stereotyped, and (2) with limited resources available, agricultural development program officials have generally put most of these resources into field staff and failed to start with a comparative cost-effectiveness analysis involving alternative methods.

There are two questions that diffusion-adoption research seems to have overlooked in evaluating channels of communication: the purposes for which various channels are intended when they are built into the strategy, and the quality and fit of the material to be used. For example, if radio were used specifically to create "awareness" and the material was boring and broadcast at inappropriate times (which Fraser [1983] notes is often the case, diffusion researchers would be unlikely to discover much radio influence at the "adoption" stage.

Another legitimate criticism of diffusion research is that it accepted the myth of the passive peasant, and thus puts too much emphasis on communication as the key variable, which in turn leads to a preoccupation with the quantity and quality of agricultural extension agents (Whyte and Boynton 1983; Saint and Coward 1977).

The Training and Visit (T&V) System championed by Daniel Benor (Benor and Baxter 1984) and promoted widely by the World Bank takes the most extreme position in selecting extension methods and media. It is built exclusively on face-to-face interpersonal contacts with *no* role assigned to other kinds of channels (Colle 1984). However, some more recent adaptations of the T&V approach in India, especially in the state of Uttar Pradesh, include communication support units that incorporate audio-visual and mass media in the extension effort.

Contemporary Communication Problems

As we examine changes in the characteristics of the farmer clientele for agricultural technology, changes in available human and physical resources, and changes in the nature and penetration of a new and traditional communication media, we need to question the appropriateness of the conventional extension strategy for communicating science information.

Extension Workers as Fragile Links in the Communication System

To deal with the question of how to communicate scientific information to farmers effectively, it is important to recognize some of the realities in this conventional linkage. Again, we need a trip to the field. Excerpts from a report of a two-year research survey of communication in the Caribbean banana industry provide a picture.

> Of 22 practices recommended to banana growers by these Associations the survey found that only four were adopted by 90% or more of the farmers in all islands. From among these 22 practices WINBAN (West Indies Bananas) and the project research staff selected for the research survey a short list of the six "most important practices." When the survey started, not one of these had been adopted by 90% or more of farmers in any island.
>
> First, it is necessary to examine the way information gets to the farmer. Information goes from WINBAN Research to the Ministry of Agriculture, the Banana Association and their field staff and then, via personal visits, farm demonstrators, printed materials and radio programmes, to the farmer. At every stage there are impediments. Partly as a result, when asked whom they relied on for most of their information, 44% of farmers said themselves.
>
> One reason for the inadequate contact between these agencies and growers is the lack of staff. The ratio of extension officers to farmers ranges from 1:122 to 1:167, and not all of even these few people can devote all their time to banana growers.
>
> Another reason is the officers' lack of knowledge. The survey indicates that the field staff possess correct knowledge of fewer than 10% of WINBAN recommendations. Indeed, they gave many recommendations which contradicted WINBAN's. This problem is compounded by their lack of training in communication techniques, which leads to information being ineffectively presented and the farmers failing to understand both the implications and the necessary skills required for new practices. Less than 23% of farmers reported having ever attended a banana demonstration, which reflects the field staff's inability to mobilize social groups and to conduct

demonstrations in an interesting yet educational manner. (WINBAN 1976)

Similar indictments are made of radio broadcasts, where effectiveness was limited by the inappropriate language, inconvenience of program times, and some farmers' lack of accessibility to a radio set.

In the Caribbean and elsewhere, evidence suggests that an extraordinary responsibility has been placed on a fragile linkage system. Coombs (1974) notes that throughout the developing world extension services are a poor match for the enormous tasks they will be called upon to perform in coming years. Based on research in fifteen countries, he concludes that most extension services

- Go it alone, with insufficient cooperation with complementary services
- Operate haphazardly with neither priorities nor plans
- Spread themselves too thinly to be effective
- Concentrate their efforts on larger producers and major commercial crops while neglecting smaller farmers and the local crops of key importance to subsistence families
- Spend little effort diagnosing the differing needs of their client farmers, and instead hand out standardized recommendations that many of these clients find impractical and useless
- Depress the productivity of their field agents by neglecting their in-service and refresher training, burdening them with distracting chores, providing them with inadequate transport, and failing to reinforce them with mass media and other communication support.

The New Clientele for Research-Based Knowledge

To this picture we must add the fact that political, economic, and social realities of the 1980s are bound to compound such problems. In their examination of the challenges and strategies "to feed this world," Wortman and Cummings (1978), who were long associated with agricultural research activities and the Rockefeller Foundation, concede that until recently there have been few direct attempts to help large numbers of small farmers increase productivity and incomes, "with ***numbers*** of farmers to be benefited as the goal" (emphasis added). Extension services may have as few as fifty staff members to serve a million farming units (Rogers 1982). And there seems little prospect that the projected number of trained personnel graduating annually from existing agricultural institutions can satisfy as much as half the potential demand. No developing country produces enough extension workers (McNamara 1973), and this will continue to be true if the labor-intensive interpersonal model prevails.

It is not numbers alone that provide the communication challenge. There is a broader range of types of clientele with whom field workers must communicate. Shahhat, as reported earlier, is one of those newly discovered farmers whose literacy, motivation, and traditions complicate efforts to maximize the impact of research information. It is documented elsewhere in this book (Sachs and Caye) that women play a key role in agricultural production, vital post-harvest activities, and marketing, but that they have traditionally not been in the mainstream of recipients of agricultural services, including attention from extension workers. For example, in the Solomon Islands, women are directly involved in growing crops for local consumption, but the Department of Extension has not resolved the problem of how to reach them.

Part of the problem lies in the lack of female extensionists. Some 20 percent of the 300,000 extension workers around the world are female, and 41 percent of these are in programs related to home economics (Weidemann 1985). Training programs for extension workers such as those at the Agricultural Training Institutes in Bangladesh have recognized a "gender gap" in the extension field staff and have begun to introduce courses to train women for extension work among women agriculturalists.

There is another side to the scientist-consumer communication issue which offers a marked contrast to the focus on low-income farmers. Singh (1976) observes that the availability of a new, highly productive, and remunerative agricultural technology has produced another "new kind of farmer."

> They have a broader orientation to the outside scientific world, and more hope, belief, trust, and respect for scientific methods of agriculture, and for testing, and adopting new ideas and practices. Further, their information needs are becoming more precise and specialized.

Decisionmaking for them is more crucial than ever before, and they depend heavily on correct and complete information about new technology. It is not unusual, for example, for a scientifically sophisticated Guatemalan rubber grower to send a son to Malaysia to learn about the latest rubber technology for incorporation into Guatemalan methods.

Extension workers' information-packaging problems become more extreme as the scope of their clientele broadens. Awa (1981) suggests that extension workers' effectiveness as "linkers" depends on their ability to convey the sense of a technical report in plainer words, "bringing the abstract into the client's frame of reference." But, he adds, in low technology nations, scientific information is frequently translated first into plain English, and then into a local language or dialect, introducing the problem of "lexical and contextual equivalence," as well as translation fidelity. Awa concludes that, to communicate research information effectively, workers must have patience, empathy with clients, sensitivity to their cultural and linguistic world, as well as familiarity with the scientist's world. But "inherent defects" in Third World extension systems preclude widespread success in this task.

Scientists as Communicators

Carl Sagan is a highly respected scientist in the field of astronomy and he is known to a broad segment of the American public. In addition to his professional writing, Sagan has written for the general population (Sagan 1980). He actively participated in production and presentation of scientific information on a U.S. public television network series, and he has appeared with considerable frequency on popular late-night television talk shows. He obviously is motivated to communicate and is rewarded for doing so. In addition, he operates in a system that allows, encourages, and supports his activities.

Most agricultural scientists operate in an entirely different climate. "Scientists have made an enormous contribution with their research," says Streeter (n.d.), "but whether it was actually getting used by small farmers was not their main concern." Even where there is interest, the mechanisms for moving information into the most suitable channels usually have not been found in the research institution itself. An agricultural researcher in the Solomon Islands once conceded that "there is a lot of data lying around, but no one has time to sort it and package it." Some of the reluctance is due to the researchers' hesitation to release information that they judge to be premature, or that they fear may be distorted by those who intend to simplify it.

Rao (1972) has identified another problem regarding communication and scientists that appears to be more widespread than the Indian location he specifically describes. Regional research stations, he suggests, could act as "linking pins" between farmers and researchers by acquainting themselves with local farmers' requirements or in developing "suitable know-how." But they do not, and consequently they and their parent institutes are out of touch with local farmers' requirements. According to Rao, even though the researchers seem to be aware of their gaps in knowledge due to communication lapses, they do not feel obliged to correct them. Only about 20 percent of the projects undertaken by Indian agricultural sciences in eight research institutes originated as a result of "inward channels" from extension agents or farmers. Some of the inefficiency in communication results from having different sets of organizations and having to send requests and results back and forth through "proper channels," or a bureaucratic labyrinth.

The initiation of India's High Yield Varieties program had to overcome some of this conservatism. The minister of agriculture, C. Subramaniam, recalls the situation:

> When it came to a question of forcing the pace of the introduction of the new varieties after their tested release, there was considerable flutter in scientific and other dovecotes. "This has never been done before," went the argument, "and so it should not be tried now." When we consider the extent to which modern agriculture in many countries depends on introduction of once-exotic plants—potato, tobacco, tea, coffee, rubber—it seems all the more surprising that there was so much resistance within the scientific community itself. Those scientists . . . who pleaded for a bold step forward, were targets of intensive and almost abusive criticisms by proponents of tradition, both lay and academic. (Wortman and Cummings 1978)

Norman Borlaug once commented that "the most conservative man in traditional agriculture is the scientist, and sometimes I am not proud to be one of them" (Wortman and Cummings 1978).

Thus, this part of the agricultural system — where knowledge begins — may, itself, impede the speedy application of research information. It is against this background of scientists' ambivalence and indifference to sharing knowledge that the role of the scientist as a communicator needs to be considered.

The Problem of Goals and Communication

There is sometimes ambiguity about what policy goals drive the research process and the communication of scientific information. Researchers may be most concerned with contributing to knowledge; technocrats (including economists, planners, and agricultural officials) may be most concerned with developing the most appropriate "packages of practices" which, when applied by farmers, will result in higher yields, leading in turn to greater exports and a more favorable balance of payments. In this kind of situation, the challenge is to move information from the "knowledge centers" (usually the research lab and experiment stations) to the people on the farms. Thus, assuming the knowledge is available or is being generated, we expect to increase production by bolstering its transmission. But then comes the plaintive question: "Why don't they [the farmers] follow our recommendations?" Blanckenburg (1984a) suggests that the more relevant question might come from the farmer: "Why are the recommendations given to us not designed for our farming conditions?"

Obviously, communication issues and goal issues are intertwined. How congruent are researchers', officials', and farmers' goals? To deal with the question of communication, we need to be clear as to the nature of the discrepancies. This suggests the need to establish dependable communication links from the farm population. Furthermore, where there is a gap, whose goals will prevail? If national production goals have priority, the problem is more complex than matters of message format, language, style, and media. When goals differ, part of the communication effort will have to deal with exchanges of information, and these may be political matters rather than scientific matters. An important emphasis in the communication effort will be on orientation and persuasion along with the "what, why, and how to" issues characteristic of technical communications. Antonio's case is worth considering.

The Happy Farmer

The setting is in Latin America. The profile is drawn from real life by Gordon Sabine (1974).

Antonio is a 60-year-old farmer in Colombia. With his wife, they own sixteen acres, two-thirds of which are planted in corn. Because of some pests that he does not know how to combat, a shortage of water, and little use of fertilizer, Antonio and his wife, Mercedes, produce about one-third the output of the "usual mountainside campesino" and only one-seventh the average rate of production in a nearby valley.

> But Antonio and his family are self-sufficient. They also grow a little coffee, and cassava, and some sugar cane, and some onions, and there are chickens and a still-mooching calf, plus a cow that gives four bottles of milk a day—when Antonio can get there first.
>
> Antonio stands only some 58 inches tall and weighs only about 100 pounds. His life is work, all day, every day, from some time before

> sun-up. He reads nothing. He sees no television. He sees no movies. He has no mailbox because there are no letters. He belongs to no organizations other than the church. In the course of a week, he sees and talks to few outside his family. . . .
>
> In many ways, Antonio is unusual. He has had two years of school himself. All three children are in school, the eldest in the fifth grade. He wants them to continue, even to high school. He and Mercedes think learning is good.
>
> Antonio takes good care of his family, and if the cow is generous, there may even be enough extra milk to sell so he can buy some meat every now and then.
>
> And so what if it does require 16 acres to feed one male adult, one female adult, three children? So what if he may be producing only one bag of corn where the best technology could make 21 grow?
>
> You cannot fairly accuse Antonio of being selfish or lazy; he grows no more because he has enough, because his family does not live with hunger, because no one ever told him how hungry so many others are, because no one ever suggested he grow more.
>
> By his own hard work, he has proven he can get along without the rest of the world. He does not need us. (Sabine 1974)

If the world were made up mostly of Antonios and if the prevailing philosophy were to let the "happy farmer" rest content, there would be little hope even with the most creative packaging of scientific information.

Approaches to Improving Communication in the Eighties

The past decade has seen evidence of attempts to reduce the geographic and professional gap between farmers and researchers, so that goals are more in harmony and two-way communication links are shorter. Farming systems research (Whyte and Boynton 1983), closer cooperation between farmers and researchers in on-farm research, and creative use of paraprofessionals such as farmer-foremen, promoters, model farmers and demonstration farmers (Esman 1983) are institutional structures designed to speed research results to the fields. Agricultural officials and researchers are beginning to acknowledge the importance of indigenous knowledge (see the chapters by Compton and Warren in this book) and decentralized diffusion systems (Rogers 1982) in agricultural development, and both are likely to improve the communication of science-based information because of increased co-orientation of knowledge producers and users.

While recognizing the need to have two-way communication between farmer and researcher, it seems likely that the need to transmit information to a large and broadly heterogeneous clientele from conventional research and development agencies will continue as we push toward the end of the decade. In the next section, we examine other changes that may strengthen efforts to disseminate scientific information more effectively.

Development and Spread of Communication Media

In development programs the use of communication media has had a checkered career. Poor production, poor timing, and little convincing evaluation research has been partly responsible for the limited and subsidiary role media have played in agricultural development. Fraser (1983) reports that a UN Food and Agriculture Organization (FAO) consultant visited four African countries to study the possibility of media campaigns to help reduce post-harvest food losses, which probably account for 12 percent of all grain grown in Africa. The prospects were dimmed by the discovery in one country that although there was a well-developed broadcasting service and 90 percent of the population lived in the rural areas, substantially less than 2 percent of air time was given to rural broadcasting. Fraser contrasts this with an incident in the People's Republic of the Congo to illustrate how powerful radio can be in stimulating agricultural change. Harvesting cassava is exclusively women's work in central Africa, but when a crisis made it imperative that cassava be harvested immediately by whatever means, radio was used to suggest that men should help. According to Fraser, reports soon began to arrive that something remarkable was happening in the villages: "For the first time in memory, the men were helping the women dig up cassava. They were doing so 'because the radio said to'" (Fraser 1983).

About twenty years ago, Schramm (1964) wrote a book in which he discussed the various roles the mass media could play in national development. His comments about their potential and their limitations apply to the issue of maximizing use of scientific research.

> We have said that the media can help in education and training, and have thus distinguished what they do in these fields from what they can do in imparting information. . . . In community adult education, discussion groups and local field staffs have contributed greatly to the effectiveness of radio. Two-way communication is needed somewhere in the process—someone to guide, react, answer questions, discuss.

> The combination of mass media and interpersonal instruction—expert instruction from the media and two-way interaction with a teacher—are extraordinarily powerful.
>
> This implies, of course, that there are teachers, technical instructors, and field staffs to cooperate in the teaching process. When they are not available, or not able to teach a certain topic, then the media can and do carry a greater share of the responsibility. . . .
>
> It may well be that in the atmosphere of need in a developing country, the media can carry a much larger share of the teaching load than they have been asked to carry in economically better developed countries. We now need more field trials in developing regions to tell us just how far we can rely on them.

Such field trials were started in the mid-1980s. A highly ambitious nine-nation effort to develop generalizations about communicating science in developing nations was launched. It was called Communications for Technology Transfer in Agriculture and was funded by the U.S. Agency for International Development (USAID). Researchers started analyzing farmers' behavior, vocabulary, their cultural and economic concepts, and community social and organizational networks in order to design and test strategies that use personal contact and mass media to transfer under-employed agricultural technologies. Equally significant is the objective of drawing on all nine studies to develop guidelines for designing communication strategies that take local conditions into account. Cornell University, the University of Wisconsin, Iowa State University, and Virginia State University became collaborators with the Academy for Educational Development and several other private sector organizations in the agricultural communication research project.

Recent changes in the character of some media forms should prompt reconsideration of the conventional techniques used to link farmers and researchers. These include greater saturation and simplified media.

Greater saturation. Radio broadcasting sets are widely distributed in rural villages, and although centralized broadcast stations make localizing somewhat difficult, many nations are expanding and decentralizing their broadcast systems. There is evidence of this in the South Pacific nations, and in Egypt, India, Bangladesh, Indonesia, and other places. (The pattern in much of Latin America is one of private and locally owned stations.)

Even television is appearing more regularly in the rural communities of the Third World. In its third decade of operation,

Doordarshan, the Indian government television service, has put 180 transmitters into service, many of them low-power facilities located in rural areas, covering 70 percent of the population. In addition, some remote areas receive regular television service directly from an Indian communication satellite.

Simplified media. Though some communication technology has grown more sophisticated (color television, satellites, computers), there is a countertrend toward simpler technology. For example, video is used widely, not only by media specialists, but also by unskilled people who rather quickly learn to handle the equipment. The Canadian Film Board's "Challenge for Change" program pioneered the involvement of villagers in video productions designed to motivate and generate self-help projects among rural people in Newfoundland (Snowden and Williamson 1984). Video has been used in agricultural development in Bangladesh, The Gambia, Sri Lanka, Tunisia, Upper Volta (now known as Burkina Faso), Peru, Venezuela, and in Chile where thousands of cooperative members were trained (Dubey and Bhanj 1984). An FAO/UNDP project that used video to train farmers in Peru has spread to Mexico, Honduras, Paraguay, and Brazil. FAO is now committed to introducing this approach throughout Latin America (Academy for Educational Development 1985).

By the mid-eighties, videocassette players had began to appear in Third World towns and villages. In India, there are two video showing places for every conventional film outlet. Some buses are equipped with videotape playback facilities to entertain passengers. Portable videotape systems are appearing in agricultural research stations, or in their budgets.

Although audio tape recording technology has been available since mid-century, the development of audiocassette technology has offered a new dimension to communication by tape. Its simplicity, low cost, and portability have made it an excellent alternative to radio broadcasting for conveying information to those who cannot read. Cassette technology has also allowed indigenous production of messages, and feedback channels from farmers' groups to research and development officials (Adhikarya and Colle, 1983; Colle, Terzuola and Colle, 1975).

Other "low technology" equipment in printing and photography suggest new options for the use of media in agricultural communication.

Implications from Communication Research

In addition to the availability of new technology, more sophisticated research is being done on the uses of communication media. Ten years ago, USAID collaborated with the Guatemalan government's Ministry of Education to test the impact of radio in agricultural development. Unlike many diffusion-adoption studies, radio broadcasting was not considered a supporting medium, but rather the principal medium for communicating agricultural research information and promoting new agricultural practices. A broadcast station was constructed to serve villagers in the Quezada Valley, and the project personnel operated the radio station so there was full control over quality and timing of broadcasts. In addition to "open broadcasting," small farmer listener groups were organized to listen to cassette recordings about agricultural technology and credit. A report summarizes the results:

> Contrary to expectation, and probably because programming is so carefully tailored to local needs, radio alone seems to be having a significant impact on farmers' behavior. This particular experimental design, some say, has created an extended personal communications system rather than a traditional impersonal broadcasting sphere. Also, the monitors and agronomists appear to reinforce the radio messages effectively. (Clearinghouse 1977)

A World Bank report (Perraton et al. 1983) looks at the use of media from an economic perspective. Citing a study done in Malawi on the respective costs of the different channels of information available to farmers (extension services, radio, printed matter, training courses, and film and puppet shows delivered by mobile teams traveling from village to village), radio was found to be the least expensive channel for the amount of information received, and the extension services by far the most expensive. The report hastens to add that extension services are still worthwhile, "since certain types of information can probably only be disseminated and understood through such a system, particularly where the farmers are illiterate."

The Bank report further notes that few studies have been made to try to establish what is the best combination of media, or what type of information is best suited to each medium. "It seems probable that, on the average, one hour of listening to the radio is less effective than a one-hour visit with an extension agent. But," the report continues, "if the cost of the extension hour is 1,000 times that of an hour listening to the radio, the cost efficiency ratio for radio is likely to be extremely favorable."

> These means are only less costly, however, if the necessary infrastructure already exists and if the agricultural information programs they carry only have to bear the marginal cost of the technology involved. In our opinion, radio is an ideal *complement* to extension services. (Orivel 1983, emphasis added)

Some World Bank Initiatives

Another World Bank staff paper (Perrett 1982) has looked directly at the Bank's experience in using communication support in various sectoral projects. Perrett notes that Bank support for agricultural development has focused on strengthening and expanding the "orthodox" agricultural extension system (including the Training and Visit system). However, growing concern over the recurrent costs involved and results from the Bank's cost-effectiveness research has led to a new interest in communication media, particularly broadcast media. Where such interest exists, Perrett says, it usually relates to the supplementary role the media could perform—namely, (1) to provide farmers with essential information, (2) to provide training and information to extension personnel, and (3) to improve feedback from the field for additional research and future extension training programs. There is also some potential for using media to improve lateral communications, such as between different farmer groups and extension workers.

Other International Programs

Perhaps one of the biggest obstacles to linking researchers and farmers through media is the availability of skilled people to produce the "software." Fraser (1983) points out that one of the reasons that rural broadcasting is so dull and ineffective in most developing countries is that the media people have not been trained in creative broadcasting. The recent establishment of the International Programme for the Development of Communication (IPDC) under the auspices of the United Nations provides funding to help developing nations train people and professionalize their media. IPDC's objectives are

1. To identify the needs and priority areas for the information and communication development plans in developing countries

2. To promote the creation or extension of infrastructures for the different communication sectors as well as improved international exchange of information

3. To promote viable regional communication institutions which may play an extensive role in the planning and execution of regional projects within the programme

4. To sensitize all parties concerned as to the important role played by communication in the development process

While one cannot be sure just how newly developed "communication infrastructures" will be used, the balance of the decade will see nations gaining a greater capacity to use these resources for diffusing agricultural technology if their use is given a high national priority in agricultural planning.

Institution Building and Agricultural Communication in India

In 1969, agricultural leaders in India initiated a program in Uttar Pradesh which now seems likely to influence the communication of the scientific knowledge emerging from India's agricultural universities. This program could serve as a model for other nations. With support from the Ford Foundation, USAID, and staff from Cornell University and the University of Illinois, the G. B. Pant University of Agriculture and Technology created a "Communication Centre" that was separate from the University's Extension Department which had traditionally handled agricultural communication. The objectives of the Centre included:

1. Working with scientists to get research published, especially in channels where it could be used

2. Developing a strong communication service, staffed by qualified communicators, to interpret the results of scientific research and assist in the process of dissemination

3. Offering courses in written, visual, and oral communication at the undergraduate and postgraduate levels to provide a source of competent agriculturally sensitive communication personnel for the mass media, universities, government agencies, and agroindustries (a graduate minor was proposed for science majors to help those students improve their abilities to communicate their scientific knowledge)

4. Exercising leadership and initiating action for in-service communication education of faculty/researchers, agricultural information workers, and extension personnel at state and district levels

5. Conducting research to help guide action programs and become a major source of communication research knowledge applicable to such problems as effective communication in agriculture.

A decade later, with assistance from the United Nations Development Programme, the University created its Advanced Centre for Post-Graduate Education and Research in Agricultural Communication, "realizing that a basic problem in India is the transfer of results of Research and Technology to the greatest number of farmers, in a form in which it can be economically utilized to increase production" (UNDP 1981).

The climax of this institution-building effort was the nomination of the G. B. Pant University of Agriculture and Technology as a "centre of excellence" for agricultural communication. Like twelve other Indian agricultural universities and institutes, it is expected to become the prime resource in its field, and to provide leadership and assistance to similar institutions. Thus, India has a mechanism in place for improving the communication of scientific information. It will train scientists to communicate, assist them in packaging their information for appropriate clientele, train extension agents, media people, and other links to farmers in the information exchange process, and conduct research on the agricultural communication process itself.

Communication through the Agricultural Information Network

Another promising approach to stimulating the flow of information between research stations and users was launched in the South Pacific in the early eighties. While the conditions are unique, the system that was created can be applied elsewhere. The South Pacific Region Agricultural Development Program (SPRAD) was funded by USAID and implemented by the University of Hawaii, Cornell University, and the University of the South Pacific (USP). SPRAD's objective is to strengthen USP's agricultural teaching, research, and extension programs, and thereby promote agricultural development in the region. The USP serves an eleven-country constituency, spread over thousands of square miles of ocean, with nations ranging from 2000 (Tokelau) to 550,000 people (Fiji).

Great distances over water make transportation and communication a major challenge. However, each country was linked into a satellite communication network, which the USP used for administration and extension teaching. The satellite also created a unique Agricultural Information Network (AIN), which was based at USP's School of Agriculture in Western Samoa.

The original project document tended to emphasize a somewhat conventional agricultural extension function: getting information out from agricultural researchers to users "on demand." However, two factors led to a modification of the plan.

1. It was recognized that a substantial amount of research information already existed in laboratories and research stations throughout the region's ten agricultural countries. (The eleventh had no agricultural activity.)
2. Information received in a country often went to file drawers or library shelves with little opportunity for use by researchers, extension workers, or agriculture teachers who didn't know it existed.

Thus, the AIN assigned to most of the constituent nations an "agricultural liaison officer" (ALO) whose responsibility it was to collect and market research information. The person's responsibilities included ferreting out research findings and reporting them to the AIN headquarters in Western Samoa, where they could be "packaged" for distribution via communication satellite or mail pouch throughout the region. The ALO was also responsible for receiving information from the network and aggressively seeking out the person to whom the information had particular relevance, such as a local scientist, the extension service, or a local radio station.

By establishing a new ALO role in its agricultural development program and providing appropriate training for the incumbent, SPRAD created an infrastructure that would stimulate the flow of communication among various participants in the agricultural enterprise.

Communication Planning

A review of the diffusion research and related communication developments over the past decade or so shows that there are few specific formulas for simplifying, packaging, or transmitting scientific information, or for setting up two-way communication links between researchers and farmers. Considerable variability in cultures, values, linguistic patterns, and visual literacy suggest caution in establishing rules. For example, research and practice offer few possibilities for universal guidelines in designing visuals for nonliterates in rural areas (Brown and Colle 1986). However, one of the most important developments to appear in recent years that will substantially contribute to effective communication of research findings is the growing attention given to communication planning and its close conceptual partner, communication strategy. Communication planning involves carefully analyzing the communication problem, laying out a strategy, identifying and allocating resources, designing the implementation scheme, and laying out a system for evaluation. A strategy is valuable in efforts to communicate science because it

- Provides a blueprint for action, showing where the communication sector is supposed to be heading and when
- Helps officials examine and plan for resource needs
- Encourages them to set priorities, because everything that might be done cannot be done at the same time, especially with limited resources
- Promotes coordination within the communication component, and among sectors

The planning and strategy approach appears in a recent manual, *Communication Strategies: A Guide for Agricultural Change Agents* (Lionberger and Gwin 1982). It also appears in a source book, *Communication Strategies for Productivity Improvement* (Rosario-Braid 1979), designed for Asian readers, including planners, communication managers, and entrepreneurs in government and nongovernment agencies engaged in industry and rural development. The planning and strategy approach is evident in the social marketing methods used in various rural development projects (Fox and Kotler 1980), including the Communications for Technology Transfer in Agriculture research project discussed earlier.

The importance of designing strategies for communicating science is that the process encourages an agency to analyze alternative approaches more systematically than is usually the case. The process starts by clearly expressing communication objectives, an exercise that may require data gathering and analysis. It includes identifying publics (or "audiences"), and matching them with objectives. Specification of the content and selection of media are important steps in building a strategy, and these must be matched with the publics and the objectives, and put into a time frame. Priorities must be set according to the resources required and available. Since choices from among alternatives (e.g., media vs. change agent) cannot often be rationally made without additional information, a strategy demands that an effort be put into data gathering: data about people's culture, demographics, attitudes, behaviors; data about media use, media strengths and weaknesses, and costs; extension agents' capabilities; agricultural input availability; data about user constraints, user opinion, leaders' attitudes; etc.(Colle 1983).

It is the data gathering associated with developing the communication strategy that provides the clue as to what and how much needs to be done to simplify or translate language or use pictures; whether to use extension agents or radio, or a combination; and how much two-way communication needs to be built into the system. The analysis helps one escape from the prison of tradition and convention and examine a broader range of options.

In summary, we may not yet have knowledge about universally applicable techniques for message design, but we have social science concepts and research methods available to develop the strategies for communicating scientific information according to the circumstances encountered.

Communicating with Shahhat

We do not yet know how best to communicate with Shahhat, partly because research has not yet provided the answers. A World Bank research team (Perraton et al. 1983) asserts that there are still questions that have not been properly clarified: the question of the interaction between the educational level of farmers and their access to extension services, and that of the best communication vehicles for the various types of information; and the question of the best vehicle in terms of cost and effectiveness for each type of information. The fact that these questions are beginning to be asked bodes well for the process of communicating science in the 1980s. Not only are we exploring institutional structures and developing training potential for making communication of science more effective, we are also ***making communicating science more scientific.***

References

Academy for Educational Development. *Beyond the Flip Chart.* Washington, D.C.: Clearinghouse on Development Communication, 1985.

Adhikarya, Ronnie, and Royal D. Colle. *Reaching Out: The Role of Audio Cassette Communication in Rural Development.* Singapore: Asian Mass Communication Research and Information Centre, 1983.

Awa, Njoku E. "Transplanting the U. S. Extension Model into Low-Technology Nations: Some Cultural and Structural Problems." Paper prepared for the annual meeting of the International Communication Association, Minneapolis, 1981.

Bajat, S. S., and S. S. Thorat. "Flow of Agricultural Communication." *Communicator* 13, Nos. 3, 4, (October 1978).

Benor, Daniel, and Michael Baxter. *Training and Visit Extension.* Washington, D.C.: World Bank, 1984.

Blanckenburg, Peter von. *Agricultural Extension Systems in Some Asian and African Countries*. Rome: Food and Agriculture Organization, 1984.

——. "Why Don't They Follow Our Recommendations?" *Quarterly Journal of International Agriculture* 23, No. 3 (1984a).

Brown, Sandra, and Royal D. Colle. "Pictorial Conventions That May Hinder Communication Efforts in Developing Countries." *Media in Education and Development*, 19, No. 4 (December 1986).

Cernea, Michael M., John Coulter, and John F. A. Russell. *Agricultural Extension by Training and Visit: the Asian Experience*. Washington, D.C.: World Bank, 1986.

Clearinghouse on Development Communication. *The Basic Village Education Project*. Washington, D.C.: Academy for Educational Development, 1977.

Colle, Royal D. "Communication Planning for Effective Nutrition Programs." Paper prepared for the Asian Regional Workshop on Effective Communication for Nutrition in Primary Health, Bangkok, 1983.

——. "Who Moves the Farmer: Some Issues in Communication and Extension." In *World Food Issues*, 2nd ed., Ithaca, N. Y.: Center for the Analysis of World Food Issues, Cornell University, 1984.

Colle, Royal D., Robert G. Terzuola, and Susana F. de Colle. "Technology on Tape." In "Focus: Technical Cooperation." *International Development Review* 17, No. 3 (September 1975).

Coombs, Phillip H. *Attacking Rural Poverty: How Nonformal Education Can Help*. Baltimore: Johns Hopkins University Press, 1974.

Critchfield, Richard. *Shahhat, An Egyptian*. Syracuse: Syracuse University Press, 1978.

Dubey, V. K., and S. K. Bhanj. "Use of Video in Rural Development." *Interaction* 2, No. 3 (1984).

Esman, Milton J. "Paraprofessionals in Rural Development." Staff Working Paper No. 573, World Bank, Washington, D.C., 1983.

Fox, Karen F. A., and Philip Kotler. "The Marketing of Social Causes: The First 10 Years." *Journal of Marketing* 44 (Fall 1980): 24-33.

Fraser, Colin. "Adapting Communication Technology For Rural Development." *Ceres*, 16, No. 5 (September-October 1983): 23-27.

Lionberger, Herbert F., and Paul H. Gwin. *Communication Strategies: A Guide For Agricultural Change Agents*. Danville, Ill.: Interstate Printers and Publishers, 1982.

McNamara, Robert F. "Address to the Board of Governors." World Bank, Washington, D.C., 1973.

Orivel, François. "The Impact of Agricultural Extension: A Review of the Literature." In "Basic Education and Agricultural Extension," Hilary Perraton, et al. Staff Working Paper No. 564, World Bank, Washington, D.C., 1983, pp. 1-58.

Perraton, Hilary, et al. "Basic Education and Agricultural Extension." Staff Working Paper No. 564, World Bank, Washington, D.C., 1983.

Perrett, Heli E. "Using Communication Support in Projects." Staff Working Paper No. 551, World Bank, Washington, D.C., 1982.

Prasad, C. *A Study of Agricultural/Rural Extension Experiences in India*. New Delhi: Indian Council of Agricultural Research, 1985.

Rao, Radha C. "Communication Linkages in Transfer of Agricultural Technology." *Economic and Political Weekly* (December 1972): A157-70.

Ray, Howard E. "Extension and Communication Models." In *Incorporating Communication Strategies Into Agricultural Development Programs.* Washington, D.C.: Academy for Educational Development, 1985.

Rogers, Everett M. *Diffusion of Innovation*, 3rd ed., New York: The Free Press, 1982.

Rosario-Braid, Victoria. *Communication Strategies for Productivity Improvement.* Singapore: Asian Mass Communication Research and Information Centre, 1979.

Sabine, Gordon A. "We resolve to . . . ." In *Communication Strategies for Rural Development*, Robert H. Crawford and William B. Ward, eds. Ithaca, N. Y.: New York State College of Agriculture and Life Sciences, Cornell University, 1974.

Sagan, Carl. *Cosmos.* New York: Random House, 1980.

Saint, William S., and J. Walter Coward. "Agriculture and Behavioral Science: Emerging Orientations." *Science* (August 19, 1977): A733-37.

Schramm, Wilbur. *Mass Media and National Development.* Stanford: Stanford University Press, 1964.

Singh, K. N. "Communication Strategy for Transfer of Agricultural Technology." New Agricultural Technology and Communication Strategy. Bombay: National Institute of Bank Management, 1976.

Snowden, Donald, and Tony Williamson. "Community Development in Newfoundland." *Interaction* 2, No. 3 (1984): 22-27.

Streeter, Carrol. *Reaching the Developing World's Small Farmers.* New York: The Rockefeller Foundation, n.d.

United Nations Development Programme (UNDP). "Project Document: Centre for Agriculture and Rural Development, 1973." IND/77/003/A/01/12.

Weidemann, Celia Jean. "Extension Systems and Modern Farmers in Developing Countries." *Agriculture and Human Values* 2, No. 1 (Winter 1985): 56-59.

Whyte, William F., and Damon Boynton. *High Yielding Human Systems For Agriculture.* Ithaca, N.Y.: Cornell University Press, 1983.

WINBAN, "How a Lack of Communication Slows Development." *Intermedia* (August 1976): 20-21.

Wortman, Sterling, and Ralph W. Cummings, Jr. *To Feed This World.* Baltimore: Johns Hopkins University Press, 1978.

CHAPTER 4

Women in Agricultural Development

CAROLYN SACHS
VIRGINIA CAYE

Throughout the world, women are involved in agricultural production. Women prepare land, plant, weed, harvest, process, and market agricultural goods. In developing countries and in many areas of the world, they have primary responsibility for meeting the subsistence needs of their families. However, the contribution of women to agricultural production regularly goes unnoticed or is undervalued by development planners. Agricultural development efforts have overlooked women's contributions to food production and have oriented projects primarily toward men. This has altered the division of labor between women and men and has in many cases affected women adversely. As development intensifies, women's work loads have increased, the status of their work has declined, and poor women especially have experienced increased difficulty meeting the subsistence needs of their families. As a result, the status of women has become increasingly marginal, with severe consequences for societies in general. In many cases, women's ability to provide adequate food for their families has decreased. The end result has been increased malnutrition in many areas. The success or failure of agricultural development projects depends on an understanding of the differential activities of, opportunities for, resources available to, and constraints on women and men.

Women's Work in Agriculture

Beginning with Ester Boserup's ground-breaking book, the importance of gender as a factor in the division of labor in agriculture and of differences in the sexual division of labor in various areas of the world have become increasingly recognized as issues in development (Boserup 1970). Nonetheless, the extent of women's economic activity is not apparent in production statistics. Because women are engaged primarily in subsistence production, their work is usually unpaid labor and therefore not valued in economic measures. In response to the inadequate measurement of women's economic contributions in agriculture, scholars have recently documented the extent of women's work through microlevel studies. Women's work in agricultural production is becoming increasingly visible as studies reveal that women's participation in agriculture is extensive and is increasing in many countries (Chaney, Simmons, and Staudt 1979; Deere 1976; Tinker 1981). Recent United Nations estimates suggest that women farmers produce at least 50 percent of the world's food. The extent of women's work in agricultural production varies by region and within regions by class, ethnicity, and country.

Africa

Throughout Africa, women have traditionally had primary responsibility for agricultural production. In Africa, women perform approximately 70 percent of the agricultural work (UN Economic Commission 1977). Prior to colonization, women had primary responsibility for agriculture in most African societies. According to the traditional sexual division of labor, women were assigned the tasks of planting, weeding, harvesting, processing, and storing crops. Men were usually responsible for clearing and preparing the land and tending the livestock.

With the intrusion of colonialism, men's contribution to subsistence activities declined. Men were recruited to work on plantations, in mines, and on other colonial projects. The absence of men coincided with an increase in other constraints on subsistence production: declining soil fertility, intensified farming as a result of population pressure, limitation on land expansion because of white settlers, and allocation of the most fertile land to export crop production. Women worked harder and were required to find increasingly more creative methods to maintain their families (Strobel 1982). In her study of Luo women in Kenya, Hay notes that from the 1930s, women bore the burden of the transitions occurring in rural areas. Women adopted labor-saving strategies in agriculture, including planting new crops such as cassava. Cultivating the land was no longer viewed as a means to acquire wealth but, rather, as a means for a woman to meet the subsistence needs of her family (Hay 1976).

Surveys of labor inputs into agricultural production in various countries acknowledge the extent of women's contributions in various tasks. For example, a study in Zambia found that women work an average of 42 percent more time in the maize fields than men (Rogers 1979). The study notes that the crucial operation limiting maize production is weeding, which is performed by women. Spring and Hansen report that continuous cultivation in Zambia has resulted in increased weed populations and consequently the expansion of the task of weeding. Weeding has remained women's work, although labor constraints often limit women's ability to weed at the appropriate times. In Swaziland, agronomists have also identified untimely weeding as a major constraint on maize production. Women perform the weeding and are often unable to weed when necessary due to a multitude of other demands for their labor (Sachs and Roach 1984; Spring and Hansen 1979).

Although women have had increasing responsibility for subsistence production, men are not excluded from agricultural production. Colonial policies have encouraged male involvement at the village level in the production of cash crops such as cotton, cocoa, and coffee for export. As a result of official policies, men often control the earnings from the cash crops although women perform most of the labor.

Asia

The work of Asian women in agriculture differs substantially from that of African women largely because of greater class differentiation and commercialization in agriculture (Blumberg 1981). The nature and extent of women's work varies by cropping system and class. As opposed to Africa, where women work primarily as farmers or family workers, Asian women working in agriculture are more likely to be employed on plantations or as wage laborers.

In Sri Lanka, 72 percent of the female work force in agriculture are wage earners. Women constitute over half of the labor force on tea plantations in India and Sri Lanka, 44 percent of the workers on coffee estates in India and Malaysia, and 30 percent of the labor force on rubber estates (United Nations Report 1978). Both Blumberg and Tinker explain that the percentage of women working on plantations has increased partially as a result of wage differentials between women and men workers and the desire of plantation owners to secure cheaper labor (Blumberg 1981; Tinker 1981). In Sri Lanka, women earn 80 percent of men's wages for identical work.

The extent of Asian women's involvement in agriculture varies according to the availability of irrigation. In nonirrigated agricultural regions, usually only the poorest rural women work as agricultural laborers. Regions characterized by irrigated rice production require a heavy labor input for women. In irrigated rice cultivation, women are frequently responsible for planting, transplanting, weeding, and harvesting, while men are responsible for land preparation (Blumberg 1981).

The introduction of Green Revolution technologies has had different impacts on women and men, resulting in increased inequities in the population in general and for women in particular. In India, the introduction of high-yielding crop varieties, farm machinery, chemicals, and irrigation contributed to the decline of women's importance in agriculture. From 1961 to 1971, the percentage of women cultivators declined from 56 to 31 percent, while the percentage of women agricultural laborers increased from 24 percent to 50 percent over the same time period (Jacobson 1976-1977).

An exception to the correspondence between women's decreased importance and the introduction of new agricultural technologies has been noted in parts of Indonesia. In West Sumatra, where women control the land, the introduction of high-yielding rice varieties has improved women's economic position (Blumberg 1981). However, situations where women control the means of agricultural production are the exception rather than the rule.

In some regions of Asia, a factor that contributes to the invisibility of women's work in agriculture is the existence of *purdah*, the seclusion or veiling of women. The subsistence work performed by women where *purdah* exists is literally out of the sight of agricultural developers and researchers. In areas of India and Bangladesh, where a considerable degree of *purdah* is practiced, women's work in fruit and vegetable production is crucial for subsistence. Women also perform postharvest processing of rice (Rogers 1979). Although women may be secluded, they contribute substantially to agricultural production.

Latin America

The latifundia-minifundia system forms the context in which women's agricultural work occurs in Latin America. Concentration of land ownership under the latifundia-minifundia system results in severe inequities and extreme poverty in rural areas. Agricultural workers, both women and men, are either wage laborers on the latifundia or small farmers on the minifundia. The ongoing commercialization of agricultural production has altered the makeup of the agricultural work force. The resident, permanent work force on the latifundia has largely been replaced with seasonal temporary workers. This seasonal work force is composed predominantly of men (Blumberg 1981).

Men migrate from minifundia, while many women remain to provide the family food supply. Because much land has been converted to cash crops, less land is available for the peasants, who are migrating to the cities in increasing numbers. Women's importance in agriculture has increased as poverty is intensified. In a study conducted in northern Peru, Deere found that the poorest women were most likely to participate in agricultural work. Women in the poorest families in northern Peru participated in 74 percent of fifteen field tasks, while women from middle-level peasant households participated in only 62 percent of field tasks. Deere notes that as agricultural production on the minifundia becomes more marginal, women's participation increases (Deere 1979).

The World Economy and Women's Work

The positions of both women and men in Third World countries have been altered by economic and political dependence on Western nations, beginning with the onset of colonialism in the sixteenth and seventeenth centuries. Export crop production and the increasing emphasis on cash crops in developing countries has created agricultural wage labor forces and transformed indigenous agricultural systems. The division of labor between women and men in traditional agricultural activities has altered as agriculture has been reoriented toward export crop production.

As noted earlier, women have been and continue to be the primary food producers in many parts of the world. Women's work and status are defined through an interplay of indigenous societies and patterns of colonization or integration into the world system (Sachs 1983). Often women in traditional societies were subordinate to males. For example, the eldest male in the household usually controlled the labor of women and younger males and with the intrusion of colonialism, women's status seldom improved. Moreover, the relative importance of subsistence economies declined under colonial rule, thereby undermining the subordinate but recognized importance of women's roles. Colonial policies aimed at the production of cash crops led to the separation between social and domestic labor, the institution of larger-scale production, and the institutionalization of private property, thus altering the division of labor between women and men and often lowering the status of women's work.

The current relationship between women's work in agriculture and the political and economic underdevelopment in Third World countries is elaborated by Deere, who suggests on the basis of her research in Latin America that women's work in subsistence agriculture allows the male wage to be lower than is necessary to support a family. In areas where women are able to support their families through subsistence activities, export industries and plantations are able to hire men at wage levels that will not support women and children. The low wage contributes to capital accumulation in the developing countries, which is often transferred to developed countries through unequal exchange systems. Thus, capital accumulation depends in part on the unpaid work of women in both agricultural and domestic activities (Deere 1976).

Beneria and Sen take Deere's work a step further, relating changes in women's work to the form of capital accumulation in a particular region. Development generates new forms of class and gender stratification, which vary by region. Beneria and Sen delineate five possible consequences of development for women: (1) women's work load increases; (2) women lose control over productive resources and the labor process; (3) women become migrant wage laborers; (4) traditional forms of patriarchal control subside but new forms are introduced; and (5) greater class differentiation occurs among women (Beneria and Sen 1981).

Women's Domestic Activities

Women's daily activities include a variety of tasks in addition to agricultural labor. Women in Third World countries also perform nonagricultural wage employment and domestic activities. Women's responsibility for childbearing, childrearing, and domestic work continues regardless of their involvement in agricultural production or other economic activities. Numerous scholars agree that such responsibilities form the basis of women's subordination.

Women in most societies are largely responsible for providing goods and services for their families, which in subsistence societies involves a variety of labor-intensive and time-consuming activities. Gathering fuel, carrying water, washing clothes, and producing, processing, and preparing food occupy a significant portion of the rural woman's day. Women's status cannot be explained solely on the basis of colonial attitudes toward women. The transformed sexual division of labor that emerges with capitalist development varies by region as a result of the interaction between indigenous and capitalist forms of production. Both traditional and Western societies emphasize women's domestic roles, thereby failing to recognize women's role in agricultural production.

Implications of Development Policy on Women's Work in Agriculture

Development has transformed agricultural production processes and has altered the sexual division of labor in agriculture. As has been described, women's work in agriculture has changed throughout the world in various ways in response to agricultural development policies, including changes from subsistence to cash crop production, introduction of technology, changes in land tenure systems, and migration.

Food and Cash Crops

The primary focus on agricultural development has been to increase agricultural production. Recent agricultural development programs have encouraged increased production both in large-scale commercial agriculture and among subsistence farmers. Subsistence farmers are increasingly encouraged to enter commercial agriculture through cash crop production to supply the domestic and export market.

An increasing dualism in agriculture is occurring between cash crop production and subsistence production. Within the dual agricultural system, women are associated with a dwindling subsistence food crop sector, while men are associated with the production of cash crops such as cocoa, cotton, and coffee. A number of factors have contributed to male control of cash crops: the Western bias towards men as farmers in conjunction with indigenous values that support male control over resources; administrative practices that favor men in providing extension services, credit, tractor use, fertilizer, marketing, and land; and the provision to men of resources for growing cash crops, while women continue to be responsible for the family food supply. Despite this increasing dualism between subsistence and commercial production, women do in fact work in cash crops. Men may control the resources of cash crop production, but women often provide the labor. Thus, in many cases, increased cash crop production results in heavier work loads for women, who must labor to produce cash crops in addition to performing subsistence production under increasingly deteriorating economic and physical conditions.

Introduction of Technology

The introduction of new agricultural technology has changed both the agricultural production system and the related social and economic system. Technology associated with the Green Revolution, including hybrid seeds, fertilizers, pesticides, and irrigation, offered the promise of increased production for developing countries. In several regions of the world, the Green Revolution increased overall agricultural production immediately. Proponents of the Green Revolution technology were quickly faced with a number of problems relating to equity. The underlying assumption of the development planners and scientists promoting the new technologies was that increased production would result in increased food supplies as well as surplus production. Increased production benefited the wealthier farmers in certain regions, but regional and local food shortages persisted in many developing countries. And, because use of the new agricultural technology was often beyond the means of subsistence and small-scale agriculturalists, the introduction of Green Revolution technology has frequently been accompanied by the concentration of wealth and capital, regional inequities, labor displacement, dissolution of rural livelihoods, and increased reliance on imported goods to meet food needs.

As a result of problems and criticisms associated with the Green Revolution, the international agricultural research centers instituted approaches to agricultural development designed to specifically reach and meet the needs of small farmers. Primarily developed at the International Maize and Wheat Improvement Center (CIMMYT) and the International Rice Research Institute (IRRI), small farm programs have coalesced into a development approach referred to as farming systems research and extension. Waugh, Hildebrand, and Andrew describe the background and potential of the farming systems approach. (See Chapter 9 of this book.)

Green Revolution technologies have resulted in increased inequities for small farmers in general and for women in particular. Men are most often the beneficiaries of new agricultural technologies both because men have access to resources and because they are targeted as recipients of the new technologies. Women have limited access to agricultural production resources, including land, capital, credit, and extension services. Access to cash and/or credit is essential for the purchase of hybrid seed, fertilizers, and pesticides.

Women in developing countries have difficulty obtaining credit for the purchase of agricultural inputs. Throughout the world, women generally have fewer assets than men. For example, in many African countries farmers accumulate capital by owning cattle. Women's access to cattle is limited because cattle are traditionally men's property and women are in a less advantageous position to acquire cattle (Kossoudji and Mueller 1979). Schumacher et al. identify the following constraints limiting access to credit for the poor in general and women in particular: concentration of control over resources, supply allocation problems in lending institutions, limited demand for credit, high borrowing costs, lack of collateral, lack of programs responsible to women's activities, and social customs restricting women's participation in credit systems (Schumacher, Sebstad, and Buvinic 1980). One strategy that women have pursued to combat these inequities is to rely on informal borrowing systems: relatives, moneylenders, and rotating credit systems. Rotating credit groups for women exist in many areas of the world and often permit women collectively to meet their needs for credit. Although informal credit systems provide women with access to otherwise unavailable cash, the amount of cash available is limited and inconsistent. The success of numerous rotating credit systems reveals that women will save and are important resources in the mobilization of capital.

Extension services provide farmers with access to information and training related to the use of agricultural techniques. Agricultural extension staff are predominantly males who provide information to male farmers. In Zambia, improved techniques in plant breeding materials are provided almost exclusively to men farmers (Spring and Hansen 1979). Several studies in Kenya, where women have traditionally been responsible for farming, have revealed the male orientation of extension programs (Abbott 1975; Staudt 1979). Both authors point out that women farmers' limited access to extension services is a result of the interplay between class and sex. More prosperous farmers tend to be the recipients of agricultural extension information; women farmers are often poorer because of their limited access to resources. In addition, women farmers are often not viewed as serious farmers by extension personnel when husbands have migrated in search of wage employment. The household is considered to be dependent on outside sources of income, and therefore the women who remain to perform agricultural work in the absence of their husbands are less likely than men to be provided with information from extension personnel.

Efforts to improve agricultural technology are focused primarily on cash crops that are controlled by men. The introduction of new technologies frequently eases men's work loads. Men often monopolize new equipment while women continue to work with traditional tools (Boserup 1970; Mead 1976). For example, the sexual division of labor in rice production in Malaysia and maize production in southern and eastern parts of Africa has traditionally placed the responsibility of preparing the land for cultivation on men. Men's work is increasingly performed with tractors while women continue to work with hand tools. In her study of women in a Malaysian village, Strange found that men were increasingly replacing women in rice production and harvesting. Women were responsible for rice processing until new milling methods were introduced (Strange 1980). Although women's work load is eased in such a switch, disastrous consequences follow since women are no longer able to provide rice for their families.

Land Tenure

Women agricultural producers in many regions of the world do not have the means to control land. Women's access to land has been undermined by colonial policies and development programs as well as by increasing pressure on land. Traditionally the land that women farm has been under their husbands' control, although in some societies traditional inheritance patterns allocate land to women. Rather than owning their land, women frequently had land-use rights.

Colonial policies have undermined women's access to land by disrupting traditional land-tenure systems and encouraging private ownership for individual men. As a result of the disruption of communal land-tenure patterns by the Chinese in Southeast Asia, the Spanish in Latin America, and the Europeans in Africa and Asia, male household heads own most of the land (Tinker 1976). In addition, traditional systems that guaranteed land-use rights to women have been destroyed and women lack new channels for access to land. For example, in Tanzania, where female farming predominated, colonial administrators confused the male's role in land allocation with ownership and gave formal title to the land to individual men (Pala 1977).

As land becomes a scarce resource owing to population pressure and intensification of agriculture, control of land is increasimgly in the form of private ownership. The most fertile land is transferred from subsistence to cash crop production and women frequently must continue to produce subsistence goods for their famiies on increasingly marginal land. Also women's limited access to cash resources prevents them from purchasing land (Fortmann 1980; Mueller 1977).

Land redistribution programs in the form of resettlement schemes have often been based on erroneous asssumptions concerning the labor and income of women and men. In Brain's study of two rural resettlement schemes in Tanzania, women were worse off after resettlement because they had lost access to land on which they had grown crops independently of their husbands. In the resettlement areas, land and the proceeds of the land belonged to men, which eliminated women's access to independent income (Brain 1976).

Migration

Development throughout the world has resulted in an increase in households that are de facto headed by women. Current estimates suggest that between 25 and 33 percent of households in the world are headed by women, with the proportion higher in areas with extensive male or female migration (Buvinic and Youssef 1978). Households headed by women constitute a major segment of the world's poor; women are increasingly assuming the major economic responsibility for supporting their families in Latin America, Africa, and Asia. A feminization of poverty is occurring throughout the world.

As subsistence agriculture becomes less capable of providing families with an economic base, males frequently migrate in search of wage employment. In Peru, women have assumed increased responsibility for agricultural activities as men have taken paid jobs elsewhere. Eastern and southern Africa are characterized by high rates of male migration. In Kenya, only 54 percent of males work the land, compared to 85 percent of the women. Botswana, Swaziland, and Lesotho are characterized by high rates of male migration to work in South African mines. Fortmann's study of women agriculturalists in Botswana found that female-headed households lacked access to animal draft power and had significantly less labor available to perform agricultural tasks (Fortmann 1980; Rogers 1979).

The migration of males to the wage employment sector of the economy has placed tremendous burdens on women's work load and their ability to feed their families. A tendency exists for men engaged in wage employment not to contribute to the subsistence needs of the household. Men more often than women spend their incomes on consumer goods that do not contribute to family subsistence (Jain, Singh, and Chang 1980; Safilios-Rothschild 1982). Female-headed households are the creation of development trends emphasizing wage employment and the undercutting of traditional family systems and obligations. Poor women remain in the dwindling subsistence sector with heavy work loads and fewer resources to provide for their families.

Transfer of the Land-Grant System to Developing Countries

The U.S. land-grant system both contributes to and alleviates the problems of women in developing countries. Flora and Flora (see Chapter 1 of this book) trace the influence of U.S. land-grant universities on the agricultural research, education, and training institutions in developing countries. The transfer of the U.S. land-grant system to developing countries included the transfer of assumptions concerning appropriate roles for men and women in rural areas. The land-grant institutions in the United States are organized to deal with men as agricultural producers and women as household workers. Extension systems have historically provided information according to their role division, with agricultural information provided to men and home economics information to women. The initial orientation of the extension service to men as agriculturalists is described by Elsinger:

> Agricultural cooperation (extension) in its inception and early development was viewed largely as a masculine enterprise. Little, if any, effort was made to interest the feminine members of the household. Women are assumed to be directing their efforts into channels better adapted to their talents. Farm business and its resultant profit or loss was considered outside the scope of their proper concern, the home (Elsinger 1931, 5).

Home economics extension efforts directed toward rural women were part of a larger societal effort to improve and acknowledge women's work in the home. The home was juxtaposed to the competitiveness in the business world. Women were called upon to uphold morality within the home against the immoral characteristics of the profit seeking that occurred outside of the home. As a result, women on farms in the United States were viewed as homemakers or farm wives by the land-grant institutions, and their participation in U.S. agricultural production was either ignored or undervalued. Recent documentation reveals that women in the United States have been and continue to be major participants in agricultural production. For example, a national study of U.S. farm women conducted in 1980 found that 54 percent of farm women considered themselves main operators of their farms (Jones and Rosenfeld 1981; Sachs 1983).

Although the erroneous assumption that women in the United States are not involved in agricultural production has been institutionalized in the land-grant system through the separation of agriculture and home economics, during the 1950s, agricultural extension services attempted to bridge this gap through a unified approach referred to as "farm and home development" (Wilkening 1958, 187-192). More recently, an approach known as the "new home economics" has emerged to analyze women's work in the household. The new home economics approaches household production using the analytic techniques for agricultural production. Home production is viewed as an economic activity that should be included in cost-benefit analysis (Huffman 1980, 14-23). However, while some of the analytical tools are now shared between agriculture and home economics, programs continue to be sex-segregated in the United States and in developing countries.

Despite women's active participation in agricultural production in many developing countries, agricultural development projects have been directed primarily toward male farmers. Projects for women have often focused on the woman's role in the household and are usually separate from agricultural projects directed to men, In general, male extension agents teach agricultural skills to men and female home economists teach household skills to women. The result is that agricultural innovations are taught to men while women are associated with the increasingly marginal subsistence sector. Although in some countries home economists teach women animal and vegetable production for family consumption, the gardening programs are distinct from larger-scale agricultural programs, despite the fact that women are often responsible for both types of production (Chaney, Simmons, and Staudt 1979; Sachs and Roach 1984).

The application of the land-grant approach that separates women's household production from men's agricultural production and ignores women's economic contribution is inappropriate for subsistence households. The land-grant universities are not alone in their need for redirection. The worldwide failure of development programs to evaluate the economic contribution of women has gradually been recognized, and efforts have been instituted to include women in development efforts.

Beginnings of Change: the Percy Amendment and Its Impact on Women in Development

The process of directing attention specifically to women in development programs evolved slowly in the 1960s in the United Nations, the United States, and Europe. As early as 1963, for example, the Swedish International Development Authority earmarked funds for women's programs. However, it was not until 1970, when Danish economist Ester Boserup published her book *Women's Role in Economic Development,* that women's participation in development became a public and visible issue (Boserup 1970). Boserup's book stimulated a small group of development specialists in the United States to review available data on women in developing nations and subsequently to seek changes in U.S. foreign assistance programs. This group, formed within the Society of International Development, called themselves the Women in Development Committee. In so doing, they coined the acronym WID.

At this same time, concerns were surfacing in the development community and in Congress that the traditional trickle-down theory of development was not working. The trickle-down theory held that with sufficient capital investment and infrastructure, the benefits that had accrued to the wealthy through industrialization would trickle down to poor households in the form of more jobs, a tax base for social services, and a general increase in the standard of living. It was further assumed that the increased incomes of household heads would trickle over to women and children within the family. However, by the late 1960s and early 1970s, it had become apparent that neither "trickle down" nor "trickle over" were working. There was a growing disparity between the incomes of the rich and the poor; and benefits that did reach poor families did not necessarily reach family members equally. There was a need to change the focus of development from urban, industrial, and capital-intensive programs to programs meeting "basic human needs" of education, good health, water, and an adequate food supply.

Concerned about these issues, Congress passed a series of amendments to the Foreign Assistance Act in 1973 calling for new directions in aid programs. These amendments required the United States Agency for International Development (USAID) to focus on "food, nutrition, health and population, education, and human resources, and the role of science and technology to accomplish these ends." An additional amendment, supported by several national U.S. women's organizations and sponsored by Senator Charles Percy of Illinois, mandated USAID to "give particular attention to activities which tend to integrate women in the national economies of foreign countries, thus improving their status and assisting the total development effort."

The initial response to the Percy amendment in the United States and in other donor nations was the development of women-specific projects. These projects often focused on the development of leadership capabilities and managerial skills by funding host country women's bureaus, commissions, and related institutions. Most projects developed for women continued to focus on the reproductive roles of women (e.g., health and nutrition projects) rather than on their productive roles (e.g., agricultural or skills training projects).

In the United States, the concept of including women in development took the firmest hold in organizations with a tradition of domestic programs by and for women, such as the Overseas Education Fund of the League of Women Voters and the National Council of Negro Women. These organizations were encouraged by USAID to undertake or expand international programs. At the same time, a few non-governmental research and/or development organizations and several land-grant and other university research centers with a focus on women, were established. Within the development community as a whole, many organizations rhetorically espoused the value of including women in agricultural and economic development projects; however, there was little substantive change in development assistance project focus and design.

Impact of Women in Development on Agriculture Projects

Analyses of documents of the United States' primary funder of development projects, USAID, have revealed that by and large that agency has continued to exclude women from its funded programs. A study by Staudt that systematically evaluated seven types of project development papers in USAID found that "women were targeted for special attention in only a tenth of the projects" (Staudt 1979, 23). This result was substantiated in a later study by Gladwin and Staudt of agricultural program design and evaluation documents. They found in 1983, nearly ten years after the congressional mandate to include women in development projects, that "a tenth or less of projects mention women" (Gladwin and Staudt 1983,13). An earlier study by the Food and Agricultural Organization states that "the Agricultural Services Division reported activities [benefiting or involving rural women] amounting to only five percent of their total" (Rogers 1979, 85). Other reviews of donor programs indicate that where women are not specifically mentioned, they are not included.

Factors Affecting Women in Development Programs: Land-Grant Universities

The studies cited above indicate that the legislative mandate of the Percy amendment has not resulted in increased participation of women in agricultural projects funded by USAID. The necessary changes have not occurred in large part because of the institutional structure of the major implementing organization for the agency's agricultural assistance programs, the land-grant university. USAID provides three primary types of assistance to land-grant universities. The first is funding to improve the university's capacity for transferring its agricultural expertise to developing nations. Such grants, called "strengthening grants," were established by congressional legislation because of the perception that the land-grant college system of research, extension, and resident instruction would be a useful and appropriate model for improving agriculture in developing nations. These grants expand the quantity and quality of professional and institutional resources available to USAID for their agricultural projects (Long 1982). Activities funded under such grants include development of faculty language capabilities, development of a topical specialty, and development of a sensitivity toward women in their agricultural programs. More than fifty of these grants were awarded. Yet, of land-grant universities with "strengthening grants," fewer than 15 percent have strong women's components in their international programs.

USAID also provides grants directly from its Women in Development office to encourage and support land-grant university efforts to include women in their development activities. These activities have to date focused primarily on networking, organizing rosters of individuals qualified to work in development programs, and other activities that would increase both the visibility of women's concerns and also the sensitivity of agricultural project personnel to the concerns of women. Such grants were usually awarded to home economics departments, where most of the women faculty in agricultural colleges are found, rather than in agricultural production departments where international development projects are conceived and managed.

Finally, USAID provides grants to land-grant universities to provide technical assistance in implementing projects in developing nations. This kind of organizational relationship represents the largest share of funding from USAID to land-grant universities, yet no documentation is available that specifically describes women's participation as agents or beneficiaries in these programs. However, reviews of USAID project documents cited earlier suggest that women are not substantially involved in university projects in developing countries.

Several elements of the institutional structure of land-grant universities contribute to this situation: the lack of communication between home economics and agricultural faculty; the perception of research on women as an obstacle to receiving promotion and tenure; and the lack of significant numbers of women faculty in agricultural positions. Recognizing that these aspects overlap in the day-to-day activities of the university, we will for the purpose of discussion review them separately.

Lack of communication is a problem among all disciplines in colleges of agriculture; it is especially acute between home economics and production agriculture faculty. This lack of communication college-wide has evolved with the increasing specialization in academic disciplines. It is exacerbated in agricultural colleges by the perception that home economics is unrelated to production agriculture or agriculture-based social sciences. Also, as women's economic contribution to society has gained acceptance, the perceived value of homemaking roles has declined, and with it the status of home economics.

Finally, in some cases, home economics activities are dispersed throughout the university, with traditional home economics subjects found in family living departments, education colleges, and agricultural colleges. The lack of a centralized department and the disincentive of lower status accorded to home economics have been a constraint to forging linkages between agricultural and home economics faculties.

Operating within these constraints, university-based programs to bring women into agricultural projects have had very limited impact in terms of project design, implementation, and evaluation of university agricultural programs. Consequently, funding for university-specific women in development (WID) activities has in many instances not been renewed when initial grants have expired. Rather, most USAID grants for WID activities in academia are now awarded to university consortia.

It should also be noted that on some university campuses, a tension exists between women's studies programs and WID centers or programs. This tension is the result of a perception of WID research as being more applied than theoretical; from concern of faculty in women's studies programs about the researchers' interference in the lives of Third World women; and from the distancing of women's studies programs, which often have a feminist focus, from home economics departments, which tend to be more traditional.

A second factor that impedes the inclusion of women in land-grant college agricultural projects is that the choice of women as a research subject is often considered an obstacle to promotion and tenure. In fact, college administrators who award (or refuse) grants often actively discourage research on women. This becomes a problem directly affecting international agricultural programs. Although the data on women has grown in the last fifteen years, the documentation of women's participation in production agriculture needed to justify the inclusion of women in the planning and implementation of agricultural projects does not yet exist.

Further, there is yet to be developed a broad theoretical framework for research on women that would lend more credence to the resulting analyses and that could be used to plan, conduct, and evaluate research.

> Some investments must be made in research on women in the context of the same broad framework also used in other types of development research on economic, social and political issues. Until now, research on women and development has largely been the province of individual committed scholars in many countries. Little has been done to date to create the scientific competence to carry out broadly based policy-oriented research on specific problems of women and development, so that understanding gained in one region or academic discipline can be applied to the solution of problems encountered elsewhere (Barnes-McConnell and Lodwick 1983).

A third constraint to increased attention to women in agricultural projects is the lack of women faculty members, especially tenured women faculty teaching topics in agriculture or rural sociology, in agricultural colleges. At many land-grant universities, they account for fewer than 10 percent of all tenured agricultural faculty. A 1979 study of 2,051 agricultural scientists in the United States found that 4.4 percent were female (Sachs 1983). In the last few years women have been increasingly appointed to tenure track positions but their numbers are still quite limited. The result is the lack of a critical mass of women who can have some influence on the university system and who are willing to risk trying to use that influence. If all the female agricultural faculty were interested and qualified in international work, there still would be too few to have a substantial impact on the establishment of project goals, the design of technical assistance programs, and the evaluation efforts of their university. Men involved in international development must support and recognize the importance of women in development.

In addition to the structural issues affecting the integration of women into development projects is the very important fact that land-grant universities have carried to their work in Third World countries certain assumptions about the division of agricultural responsibilities in the United States. They have assumed that men are responsible for management, resource allocations, and labor inputs in agriculture when, in fact, in many developing nations—and in the United States—

women play a significant role in agricultural production; however, their contribution has often been ignored or undervalued.

Further, where women are recognized for their role in food production in development programs, agricultural activities are often undertaken that inadvertently reduce that role. In those agricultural projects that target both sexes, women's responsibilities for family health and welfare, which usually require different kinds of extension and training programs, may be ignored. The result is often disruption of traditional agricultural systems and detrimental change to women's role, rights, and benefits.

These, then, are some of the issues affecting the ability of land-grant universities to plan and implement gender-neutral agricultural programs. Some of the same issues touch the agency that funds universities: USAID. Organized into geographic, technical, and regional bureaus, USAID also has offices or "missions" in aid-recipient countries. All bureaus participate to some extent in the agency's complex system for project design and implementation. The unit with specific responsibility for women, the Office of Women in Development (WID), is located within the technical Bureau for Program and Policy Coordination. Established in 1974 in response to the Percy amendment, its purpose is to "assure that women are taken into account in project design, implementation and evaluation" (Staudt 1979, 13).

Major activities have included the introduction in project design documents of a required "impact statement," describing the effects of a proposed project on women; the assignment of responsibility to one individual in each mission for addressing the concerns of women; training of agency staff on methodologies to ensure women's participation in projects as both agents and beneficiaries; and various efforts to facilitate the participation by women staff members on project design teams. In addition, the WID office has funded special projects for women and women's components to new or existing projects.

In 1983, the WID office issued a policy paper that defined USAID goals and strategies for women and reflected a change in the emphasis of the office from a primary concern for equity to a view of women as a potential economic resource. It also called for the disaggregation of data by sex in all major agency project documents to remedy the lack of data on women in aid-recipient nations.

Specific regulations for operating this office to fulfill its mandate were not included in the legislation creating the office, and its budget has been small since it was established. The WID office has limited authority and must usually rely on informal relationships to have its influence felt. Further, the coordinator of the office is a political appointee. Thus, the style and emphasis of the office have changed with the political tenor of the times.

The reality of the WID office mandate has been to change, with few resources, a large and entrenched bureaucracy. This change has come slowly. Although the need to include women in development has been accepted in the theory and rhetoric of USAID, changes in funded projects have been limited. As noted earlier, evaluations of project documents show that most development projects do not address the concerns of women in their design or implementation. In fact, inclusion of women in project activities occurs in many cases only if there is a supportive woman involved in project design.

Many USAID field staff address only minimally the need to include women in projects simply because it is a requirement from the head office. Other USAID personnel do not recognize that there is a problem or are confused about what the problem is. In addition, U.S. foreign assistance programs change focus and direction according to the politics of the moment. Many USAID employees see inclusion of women in project activities as just another political demand that must be superficially and temporarily addressed until its political support subsides. This lack of enthusiasm within USAID limits the extent to which procedures to include women will be made a part of individual project design, implementation, and evaluation. It further means that there will be uneven attempts to ensure that land-grant universities, as implementing agencies for USAID, will address this issue.

The USAID Structure and the Integration of Women into Development Programs

As with land-grant universities, efforts to fully integrate women into USAID projects have run up against the agency's complex structure. Three inhibiting aspects of this structure are the lack of communication within USAID, the lack of status accorded to efforts to work with and for women, and the lack of women professionals in the agency.

The sheer size and complexity of USAID render communication and coordination of its geographical, technical, and regional bureaus very difficult. Within the formal process of project development, myriad documents are drafted, reviewed, and amended. There is also an informal network of professionals through which much information is conveyed. Gaining entry into the formal and informal networks of procedures and communication is a slow process. It requires individuals concerned about women in development to be in the right places and to have their own system of communication.

A second problem is the lack of status for individuals involved in the WID effort. After assignment to a position associated with women's issues or projects, an employee is often no longer perceived as experienced or skilled in other technical areas such as agriculture or nutrition. Consequently, some employees are reluctant to be associated with activities related to women and development for fear of forever limiting future job possibilities to a WID job.

Another major barrier is the lack of women in USAID who could effect changes in program design and evaluation. In 1973, when the Percy amendment was passed, only 6.4 percent of program positions were occupied by women. By 1983, although some gains had been made, women were still a distinct minority in professional positions. In some bureaus, women actually lost ground. In the regional bureaus, for instance, which hold the keys to programming in the field, women accounted for only 5.3 percent of professional staff in the Africa Bureau, 3.7 percent in Asia, 4.8 percent in Latin America, and 5.1 percent in the Near East. Increasing the number of women in positions of authority and influence is not necessarily the key to greater equity for women in USAID project programming. However, the lack of a critical mass of women makes it more difficult to increase men's sensitivity toward the integration of women into the development process.

Host Country Governments

In the late 1960s, the primary response to United Nations initiatives related to women in Third World countries was to establish national women's bureaus or women's commissions. Many more were established for the United Nations International Women's Year of 1975. Initially perceived as a means by which support could be given to increased involvement of women in development planning and programs, women's bureaus have in general been unable to meet these expectations. Problems with integrating women into development in aid-recipient countries are structural, cultural, and political. In those countries with women's bureaus, the structural hierarchy often places them in a relatively powerless ministry, or they have a low status within a ministry. Often, lack of power results in insufficient funding to carry out their responsibilities. This is doubly damaging: not only are women's bureaus themselves unable to effect meaningful policies for women, they also impede the introduction of other programs. Governments point to them as proof that they are meeting the needs of women.

As in the United States, Third World country governments are overwhelmingly dominated by men, and increasing the number of women can be problematic. In many nations, restrictive educational policies limit the number of trained women available. In the women's bureaus, staff may have been drafted simply because they are female and may lack appropriate skills in a new position. Or women's bureaus may "monopolize the best trained women in the country who would provide a valuable nucleus of women for major projects" (Rogers 1979, 103).

There is the additional problem in Third World countries of lack of support for increasing opportunities for women in principle and for the women's bureaus who might promote this principle. Support for women's bureaus may be lacking for several reasons. Limited funds result in conflict and competition between women's bureaus and local women's organizations, many of which existed before the women's bureau was established. In some cases, "the establishment of national machinery has pre-empted efforts to improve women's status by bureaucratizing initiative and neutralizing grass-roots efforts" (Staudt 1981, 350). Also, the women in women's bureaus tend to come from wealthy families; their background and training may render them less sensitive to the concerns of the poor majority. They may want to emphasize programs in home economics rather than agricultural or skills training programs.

Support for women's involvement in development programs is also withheld because of perceived association with the U.S. feminist position that has supported such activities in the United States. They see women's increased economic and social independence as damaging to families and children. Also, there is in most nations no lobby for women's rights, changes in national policies, or development programming. In a few isolated instances, women have rebelled at the grassroots level by refusing to participate in projects or demanding revisions in projects that have been detrimental to them. But there is no lobbying on program design at the national level.

Successful Women in Development Projects

Many of the successful projects for rural women have worked with women's organizations or organized women into cooperative groups. Projects in Ghana, Cameroon, Kenya, Bangladesh, and Honduras exemplify successful development initiatives. Rotating credit associations in Ghana and Cameroon have successfully provided rural women with money to improve their agricultural operations (Bernard 1987). In Kenya, women's organizations (*harambees*) have been organized to help women obtain credit and solve business problems cooperatively. A rural credit project in Bangladesh has been highly acclaimed for providing money to rural women. In Honduras, a women's poultry cooperative was successfully established by a small private organization. The women's poultry cooperative was successful in providing women access to credit and resources (Flora 1987).

As Flora points out, the small credit and income generation projects that have been designed for women often are the first signs of recognition of the importance of women's work and may become mobilizers for participation in the social and political arena. Thus, the projects, although often limited in scope, can help meet rural women's immediate needs.

Conclusions

Women's participation in agriculture throughout the world is extensive and becoming increasingly visible in part because the success of agricultural development depends on women's participation. However, much of their work continues to be undervalued or ignored. Significant changes must be made in the development of agricultural projects to achieve success.

First, inadequate documentation of women's work must be remedied through increased data collection and research on women's contributions to agriculture. Prior to the establishment of agricultural projects, information should be obtained on many aspects of women's participation in agriculture, including the sexual division of agricultural labor by task and by crop, access to land, availability of credit, access to information, use of tools and technologies, decisionmaking, extent of female-headed households, and nonagricultural demands on women's time.

Second, the documentation of rural women's work is a necessary but not sufficient condition for increasing the visibility of women and redirecting agricultural programs and projects. Research findings must be interpreted to policymakers and translated into action. As Ceboterov has suggested, research findings on women in agriculture may be used for informing, denouncing, or announcing. Research has been most effective in informing policymakers on the crucial contribution of women to agriculture. Denouncing projects and programs excluding women is less effective than announcing alternative plans that address women's needs. Creating alternatives, however, requires a thorough understanding of the constraints women face and the possibilities for overcoming these constraints (Ceboterov 1982).

Third, the sexual stratification sytems and bureaucratic structures in both the United States and in Third World countries that set limits on the possibilities for women in agriculture must be revised. In most areas of the world, men's economic advantage results in women's dependency on men. At the same time that formal institutions reinforce women's dependence on men, development has brought changes, such as male migration and men's involvement in cash crop production, that have resulted in women assuming increased responsibilities for themselves and their families. Women, especially in Africa, are independently producing their families' food supply under systems that wrongly assume women to be dependent on men. Therefore, agricultural programs need to acknowledge the economic and social shifts in activities of women and men and attempt to provide women with access to land, capital, credit, and agricultural information. The specific needs of women in agriculture must be assessed and programs should be established to facilitate women's participation.

Fourth, the bureaucratic separation of programs that affect women's lives has resulted in ad hoc programs that fail to consider the various dimensions of women's lives. Coordination of specialists in home economics, agriculture, nutrition, and family planning is essential if the multidimensional needs of women are to be addressed. Also, the separation of women's programs from other development activities has tended to make women's programs marginal.

Finally, rural women should participate in defining their needs and strategies for change. Grassroots efforts by rural women's groups are often multifaceted and should be encouraged to develop cooperative mechanisms for credit, production, and other activities.

As indicated above, the transfer of the land-grant model of agricultural development to developing countries is inappropriate without major modifications. The modifications must take into account the importance of women's work in agriculture. The research, teaching, and extension dimensions of the land-grant system can, with appropriate adjustments, collect information concerning women agriculturalists, assess the needs of women agriculturalists, and provide information to women in agriculture. Women are a potentially powerful force for economic and social change, and projects and programs for agricultural development must be reoriented to encourage their involvement. The success of agricultural development in feeding people throughout the world depends on the participation of women.

References

Abbott, Susan. "Women's Importance for Kenyan Rural Development." *Community Development Journal* 10 (1975): 179-82.

Barnes-McConnell, Pat, and Dora G. Lodwick. *Working with International Development Projects: A guide for Women in Development.* Lansing Mich.: Office of Women in Development, Michigan State University, 1983.

Beneria. Lourdes, and Gita Sen. "Accumulation, Reproduction and Women's Role in Economic Development: Boserup Revisited." *Signs: Journal of Women in Culture and Society* 7 (1981): 279-98.

Bernard, Jessie. *The Female World form a Global Perspective.* Bloomington: Indiana University Press, 1987.

Blumberg, Rae Lesser. "Females, Farming and Food: Rural Development and Women's Participation in Agricultural Production Systems. In *Invisible Farmers: Women and the Crisis in Agriculture,* Barbara Lewis, ed. Washington, D.C.: Office of Women in Development, United States Agency for International Development, 1981, 24-84.

Boserup, Ester. *Women's Role in Economic Development.* New York: St. Martin's Press, 1970.

Brain, James L. "Less Than Second-Class: Women in Rural Settlement Schemes in Tanzania." In *Women in Africa,* Nancy J. Hafkin and Edna G. Bay, eds. Stanford, Calif.: Stanford University Press, 1976, 263-82.

Buvinic, Mayra, and Nadia H. Youssef. "Women-Headed Households: The Ignored Factor in Development Planning." Washington, D.C.: Report submitted to the United States Agency for International Development, 1978.

Ceboterov, E.A. "Research on Rural Women: An International Perspective." *Resources for Feminist Research* 11 (1982): 28-32.

Chaney, Elsa, Emma Simmons, and Kathleen Staudt. "Women in Development." Paper presented to the United States Delegation to the World Congress on Agrarian Reform and Rural Development, Food and Agricultural Organization, Rome, 1979.

Deere, Carmen Diana. "Rural Women's Subsistence Production in the Capitalist Periphery." In *Review of Radical Political Economy* 8 (1976): 9-17.

——. "The Agricultural Division of Labor by Sex: Myths, Facts and Contradictions in the Northern Peruvian Sierra." Paper presented at the Latin American Studies and African Studies Association Meeting, Houston, 1979.

Elsinger, Verna. "The Women's Sphere." *Rural America* 9 (1931): 5.

Flora, Cornelia Butler. "Income Generation Projects for Rural Women." In *Rural Women and State Policy,* Carmen Diana Deere and Magdalena Leon, eds. Boulder, Colo.: Westview Press, 1987, 212-38.

Flora, Jan L., and Cornelia Butler Flora. "An Historical Perspective on Institutional Transfer." In *The Transformation of International Agricultural Research and Development,* J. Lin Compton, ed. Boulder, Colo.: Lynne Rienner Publishers, 1988.

Fortmann, Louise. "Women's Involvement in High Risk Arable Agriculture." Paper presented at the Ford Foundation Workshop on Women in Agriculture in Eastern and Southern Africa, Nairobi, Kenya, April 9-11, 1980.

Gladwin, Christina, and Kathleen Staudt. "Reaffirming the Agricultural Role of African Women in Household Economics and Rural Development." In *Proceedings of the Association of Faculties of Agriculture in Africa, Workshop on Home Economics and Rural Development.* Oct. 17-22, 1983.

Hay, Margaret Jean. "Luo Women and Economic Change During the Colonial Period." In *Women in Africa*, Nancy J. Hafkin and Edna C. Bay, eds. Stanford, Calif.: Stanford University Press, 1976, 87-110.

Huffman, Wallace. "Farm and Off-Farm Work Decisions: The Role of Human Capital." *Review of Economics and Statistics* 82 (1980): 14-23.

Jacobson, Doranne. "Indian Women in the Process of Development." *Journal of International Affairs* 30 (1976-77): 211-42.

Jain, Devaki, Nalina Singh, and Malini Chang. "India." In *Women in Asia,* Rounaq Jahan, ed. London: Minority Rights Group Reprint No. 45, 1980, 8-10.

Jones, Calvin, and Rachel Rosenfeld. *American Farm Women: Findings From a National Survey.* Chicago: National Opinion Research Center, 1981.

Kossoudji, Sherrie, and Eva Mueller. "The Economic Status of Female-Headed Households in Rural Botswana." Paper presented to the Seminar on Rural Income Distribution Survey, Gabarone, Botswana, June 26-28, 1979.

Long, Erven J. "Evaluation of the Title XII University Strengthening Programs." Paper presented to the BIFAD Meeting, Washington, D.C., June 1982.

Mead, Margaret. "A Comment on the Role of Women in Agriculture." In *Women and World Development*, Irene Tinker and Michele Bo Bramsen, eds. Washington, D.C.: Overseas Development Council, 1976, 9-11.

Mueller, Martha. "Power and Powerlessness in Lesotho." In *Women and National Development: The Complexities of Change*, Wellesley Editorial Comittee, ed. Chicago: University of Chicago Press, 1977.

Pala, Achola. "Definitions of Women and Development: An African Perspective." *Signs: Journal of Women in Culture and Society* (1977): 9-13.

Rogers, Barbara. *The Domestication of Women: Discrimination in Developing Societies.* New York: St. Martin's Press, 1979.

Sachs, Carolyn. *The Invisible Farmers: Women in Agricultural Production.* Totowa, New Jersey: Rowman and Allenheld, 1983.

——, and Christina Roach. "Women and Agricultural Production on Swazi Nation Land." Paper prepared for South-East Consortium for International Development, Washington, D.C., 1984.

Safilios-Rothschild, Constantina. "The Persistence of Women's Invisibility in Agriculture: Theoretical and Policy Lessons from Lesotho and Sierra Leone." Working Paper No. 88, Center for Policy Studies. New York: The Population Coucil, 1982.

Schumacher, Ilsa, Jennefer Sebstad, and Mayra Buvinic. *Limits to Productivity: Improving Women's Access to Technology and Credit.* Washington D.C.: International Center for Research on Women, 1980.

Spring, Anita, and Art Hansen. "Women's Agricultural Work in Rural Zambia: From Valuation to Subordination." Paper presented to African Studies Association Annual Meeting, Los Angeles, Oct. 31, 1979.

Staudt, Kathleen. "Class and Sex in the Politics of Women Farmers." *Journal of Politics* 41 (1979): 492-512.

——. "Tracing Sex Differentiation in Donor Agricultural Programs." Paper presented to the American Political Science Association Annual Meeting, Washington, D.C., Aug. 30, 1979.

——. "Women's Organizations in Rural Development." In *Invisible Farmers: Women and the Crisis in Agriculture*, Barbara Lewis, ed. Washington, D.C.: Office of Women in Development, United States Agency for International Development, 1981, 329-90.

Strange, Heather. "Some Changing Socioeconomic Roles of Village Women in Malaysia." In *Asian Women in Transition*, Sylvia Chipp and Justin Green, eds. University Park: The Pennsylvania State University Press, 1980, 123-51.

Strobel, Margaret. "African Women." *Signs: Journal of Women in Culture and Society* 8 (1982): 109-31.

Tinker, Irene. "The Adverse Impact of Development on Women." In *Women and World Development*, Irene Tinker and Michele Bo Bramsen, eds. Washington, D.C.: Overseas Development Council, 1976.

——. "Policy Strategies for Women in the 1980s." *African Report* 26 (1981): 11-16.

United Nations Economic Commission for Africa. "The Changing and Contemporary Role of Women in African Development," *Economic Bulletin for Africa* (1977).

United Nations Report of the Secretary General. "Effective Mobilization of Women in Development." New York: United Nations, 1978.

Waugh, Robert K., Peter E. Hildebrand, and Chris O. Andrew. "Farming Systems Research and Extension." In *The Transformation of International Agricultural Research and Development*, J. Lin Compton, ed. Boulder, Colo.: Lynne Rienner Publishers, 1988.

Wilkening, Eugene A. "Joint Decision-Making in Farm Families as a Function of Status and Role." *American Sociological Review* 23 (1958): 187-92.

CHAPTER 5

The Integration of Research and Extension

J. LIN COMPTON

Theodore Schultz's seminal work, *Transforming Traditional Agriculture* (1964), was appropriate for its time. World conditions today seem to suggest that we need to transform international agricultural research and development programs that have as their major goal improvements in the lives of those who practice traditional agriculture. This chapter will provide a brief look at the past, identify some of the lessons learned, assess the present state of affairs, and project what some of the likely future developments will be in the area of integrating research and extension activities.

Research and extension are two components within a matrix of marketing, supply, governance, and production. The interactions between and among these components are as critical as the interactions that take place within each one. The focus here, however, will be confined to an assessment of education and communication factors that affect the larger knowledge creation-diffusion-utilization system, or research and extension.

The following treatment encompasses a definite bias, or an overriding concern for the welfare of small, limited-resource farmers operating in multi-crop systems—the farmers who constitute the majority of the world's population. It is suggested that research and extension programs are needed to help this type of farmer make decisions among the various alternatives of a complex, mixed-cropping situation. A strong assumption throughout this chapter is that we need institutional analysis models that can assess the dynamic interaction between production systems and institutional structures. This notion guides much of the analysis that follows.

A Past Accounting

There is a strong international concern today for the problems and prospects of technology transfer. Much of that concern stems from an awareness of past failures to bring about substantial positive changes in the lives of the world's rural poor. Major and noble efforts have been made to bring the knowledge and expertise of scientists, administrators, and educators from agriculturally developed societies to bear on the problems of food-deficient nations. The Green Revolution was one such major effort. Great achievements were made in the production of grains in order to feed the world's masses. And yet, there is much doubt as to its overall impact. One analysis suggests that the Green Revolution was neither a cure for nor a cause of inequity and poverty, as many of its detractors have claimed. The lower grain prices that resulted from vast production increases have since had positive effects on income distribution. At the same time, it is safe to say that landowners have benefited more than laborers or landless poor because they have been more able to take advantage of the packages of technology developed by the revolution (Pray 1981). Another view is that the poor would, nevertheless, have been worse off in the absence of the Green Revolution.

One major lesson to be learned from past efforts to transfer technology is that such transfer itself is not and should not be the major goal. The major goal should be the transfer of science or the capacity to generate indigenous technology appropriate to the needs and circumstances of a developing society. Science here refers to the processes of scientific inquiry and discovery, not the end products developed in any one society.

The generation, creation, or discovery (research) of knowledge must be paired with a concern for its dissemination, diffusion, and use (extension). Just as we have learned that the development of agricultural research capabilities in low-income countries is a very difficult task, we are likewise coming to the realization that the use of knowledge is still a largely intuitive and underdeveloped intellectual skill around the globe. This is partly because the institutions and organizations involved in agricultural development have established reward systems that discourage interest in the application of knowledge among those who generate it. Many still hold to the adage that all we have to do is "build a better mousetrap and the world will beat a path to your door," an expression of the belief that the quality and attractiveness of an innovation will assure its adoption. There is much evidence that this is not true, and that decisions to adopt are often influenced by a very complex set of factors.

A history of the efforts behind extension programs in Third World countries shows a steady and progressive evolution in the collective understanding of how to improve agricultural practice and in the general understanding of the factors and forces that have influenced that process. During the 1960s, most agricultural extension work was guided by the theory of a diffusion of innovation. The basic assumption was that if all farmers could achieve yields as high as those of the best farmers, production would rise dramatically. The focus was on ways to persuade more farmers to adopt innovative agricultural practices and technology. Farmers were assessed in terms of their progression through the stages of an awareness of the innovation, then the development of an interest in it, a mental evaluation of the innovation, trying it out, and finally adopting or rejecting it and, if adopted, consolidating the innovation into the general farming scheme. A key strategy was to modify communication to meet the needs of various types of farmers. Major stress was placed on information and how to evaluate and apply innovation messages designed to make people more receptive to change.

Further and later analysis of the diffusion of innovation approach evinced several major weaknesses. Its proponents seemed to give little attention to circumferential factors influencing farming decisions such as credit, market, and land tenure situations. Little attention was given to the appropriateness of the technologies or innovations themselves. The tendency to focus on "opinion leaders" and "early adopters" also left out those who were most in need. One would dare say that there was a tendency for a warped view of the importance of the mass media in promoting change. This approach gave little attention to the importance of feedback as a way to change strategies, themselves, in midstream. Also, interregional and intervillage community variations received little attention from the proponents of this approach.

The social research associated with the diffusion of innovation approach also had its weaknesses. It tended to rely upon time and recall data, which is always a questionable strategy. Little research was done on collective or authority decisions affecting individual optional innovation decisions. Researchers failed to conduct relational analyses because of the overriding focus on the individual adopter.

An approach that evolved later was the package program. The general principle here was that farmers needed a combination of techniques and supports such as credit, markets, irrigation, inputs, and not just information. However, such packages tended to require a strong commitment at the national level to be successful. This was so because of the difficulty of integrating diverse components and bringing about enough interagency coordination to deliver coherent packages of technology to farmers at the times they were most needed. This approach also seemed to favor single-crop-focused systems as against multiple-crop systems, largely divorcing itself from the mass of limited-resource farmers. It was also found that despite the emphasis on upward communication flows from farmers to agencies with those agencies supposedly being responsive to farmers, there was a tendency for messages to remain highly centralized and national in character. Little effort was made to develop messages (packages of technology) tailored to the wide variations found even within one geographic area.

A later strategy has come to be called the induced innovation approach. The basic assumption of this approach is that technical changes in rural institutions can be guided along an efficient path by price signals in the market, provided that the prices efficiently reflect changes and that there exists effective interaction among farmers, public research institutions, and private agricultural supply firms. This approach stresses communication and information systems that support functioning of product and factor markets, the role of effective farmers' organizations, and the important role of administrators in manipulating factors in the general agricultural economy. Because of its focus on commercial agricultural economy, however, the induced innovation approach tends to neglect the problems and needs of subsistence farmers, landless laborers, and low-income rural people.

More recently, a wide range of decentralized models seem to be evolving that are user-oriented. Here the need of the user defines the only acceptable value stance for the change agent. The diagnosis of need is an integral part of the process of planning and conducting change programs. Outside change agents are supposed to assume a nondirective stance and peer relationship with clientele. This approach emphasizes the importance of fully using those internal resources that exist and that are accessible to the client. A major problem with this approach has been the difficulty of fitting related objectives to national goals for agricultural development—national goals that may see rural family preferences as an obstacle to be overcome rather than satisfied. Finally, the difficulties in operationalizing such an approach are legion.

An historical assessment of the various institutional structures and forms that research and extension programs have assumed or been a part of during the past three decades provides a slightly different but somewhat parallel analysis of institutional evolution. The 1950s and 1960s may be seen as a period where extension programs that were established in low-income countries reflected a direct transfer model imbued with the philosophy of the diffusion of innovation approach. Later, in some countries, an effort was made to create agricultural universities patterned on the U.S. land-grant model in order to integrate the functions of research, education, and extension. In the 1970s, attention seemed to shift to the importance of high-powered scientific approaches, such as the packaged program and induced innovation approaches. Certainly the induced innovation approach emphasized the importance of managerial science as well. More recently, attention seems to be shifting to the prospects for establishing interactive knowledge creation-diffusion-utilization systems. This is dealt with in greater depth at a later point in this chapter.

History is a good teacher. In addition to telling us what not to do, it oftentimes tells us what we still have to do. To integrate research and extension activities, we still need to find ways to relate theory with practice, basic research with applied or adaptive research, the work of scientists with the work of educators, and concern for production with concern for equity (Compton 1984a).

Lessons Learned

U.S. efforts to influence agricultural development abroad, while largely positive, have sometimes met with failure. Massive efforts in Latin America during the 1950s to establish extension programs failed because what was transferred was an organization for the diffusion of innovations, not a complete research utilization system (Rice 1974). These early efforts in Latin America would seem to be naive, in the light of our current awareness that the creation and diffusion of knowledge are as important a component as utilization in an overall system.

Great efforts were also made to implant the institution of the U.S. land-grant university into several developing countries. The major rationale behind this effort was to foster institutions that could integrate research, education, and extension. Americans generally failed to recognize that the integration of these functions in U.S. land-grant institutions in the U.S. happened more or less at the beginning. The Hatch Act and the Smith-Lever Act simply legitimized and strengthened the research and extension components of the land-grant university. In less developed countries, however, the land-grant model came as an imposition onto a network of agricultural development services that already existed. This fact has made it very difficult for university-based administrators to achieve a free hand in designing and implementing agricultural development programs.

In the United States, farmer organizations preceded the establishment of the agricultural experiment stations. To a large extent, they served as pressure groups that influenced the scope and thrust of agricultural research. Low-income countries have seldom, if ever, had significant farmer associations.

U.S. land-grant universities had an early beginning as agricultural and mechanical schools for the sons of farmers. The agricultural universities created in Third World countries were largely the work of elite groups. Upon graduation, the students of U.S. land-grant schools largely returned to occupations in agriculture. The students of similar institutions in the Third World generally assumed positions in the ranks of government civil servants.

Agriculture, in general, came to be seen as a science in the United States. There was a strong emphasis on the creation of efficiency in production. Agriculture in the low-income countries continued to be a matter of subsistence. These very different conditions in the United States and Third World countries are now clear to us. If hindsight could have been foresight, it is conceivable that institutional models could have been shaped to fit these circumstances.

Despite of the difficulty of establishing agricultural universities in developing countries, there have been some success stories. Certainly the accomplishments of India's present agricultural universities stand out. A scholarly analysis of India's experience, while pointing out some of the difficulties, illuminates some of the achievements (Nagel 1980). It would seem that while the functions of research, education, and extension are being integrated within India's agricultural universities, this integration is largely intraorganizational, rather than focusing on links with other institutions and agencies outside the university. Indian universities seem to have turned inward for self-protection. The large-scale effort to establish replicas of the U.S. land-grant institutions has served to draw Indian agricultural universities away from traditional agricultural organizations. This is ironic in that the U.S. land-grant university had to do just the opposite to protect itself. From the beginning it had to demonstrate a strong outreach and service orientation toward farmer clientele and the agencies and organizations already serving their needs.

Nagel has identified five factors influencing university research, education, and extension linkages within India's agricultural universities:

1. Concentration
2. Specificity of linkages
3. Integration
4. Practice orientations
5. Organizational efficiency in knowledge transfer

Efforts have been concentrated by pooling resources under one roof. It has been difficult for government agricultural services to achieve a comparable situation. Agricultural research and development endeavors in the university setting are more likely to focus on the most pressing problems because of the freedom to assess the nature of those problems and mobilize appropriate resources. Also, concentration of the university at the state level has made it possible to simplify communciations with field personnel and with outside knowledge centers so that accurate information regarding needs and responses is better coordinated. The professionalization of the agricultural research community occurring within the agricultural university also has the added spin-off of producing good graduates, many of whom continue to serve agriculture.

The agricultural university has also made it possible to achieve a specificity of linkages and information channels relevant for agricultural development—channels that are usually very difficult to accomplish among government bureaucracies. The agricultural universities provide a direct connection between their research scientists and their extension staff. They are centers for resource transfers and act as information clearinghouses. They also serve as a channel for research funds and for the flow of information from and to national and international centers. The separation of research and extension from administration within the university (in comparison to what happens within a government bureaucracy) allows the scientists and educators to focus on subject matter. And it has been found that as extension personnel become more involved in adaptive research, internal linkages within the university itself are improved.

The manner in which staff functions and duties are organized and managed within the agricultural university makes it possible to be well integrated and coordinated. This sometimes occurs within the same department because its staff will be involved to different degrees in two or more of the three functions. This type of dual function facilitates informal links among staff and assists the exchange of knowledge and information about agricultural problems and solutions.

It has been noted that a strong service ethic within agricultural universities orients them to activities that respond to user needs. There is some danger in becoming too responsive to clientele demands in that the university might unnecessarily encroach upon the responsibility of the state department of agriculture.

The university can also help bring organizational efficiency to the transfer of agricultural knowledge. Knowledge that gets documented seldom gets lost. But there is a general awareness that concentrated research and practice orientations are not enough to ensure a speedy transfer of this documented knowledge to clientele. A continuing active search for knowledge requires continuing support for the institutionalized research and extension teams of the university.

Other countries have not fared as well as India in their efforts to establish agriculturally oriented universities. One major problem they have encountered is how to get research and extension staff off the campus and into the field. The United States has not experienced a major problem of this nature, largely because of the multilevel and decentralized nature of the funding that is provided for extension and even for some research activities. But a focus on the integration of research and extension activities within the context of an agricultural university represents only a small percentage of the forces now operating around the world. In most countries, these functions are not the primary responsibility of universities at all, but of separate and firmly established government bureaus or agencies.

In most countries, national- and regional-level agricultural researchers answer to the capital, while extension is basically a provincial- or state-level affair. This results in a bifurcated system which inhibits the dynamic interaction between the two components needed to effectively identify, study, generate, and deliver suitable responses to farm-level problems.

Research and extension, being separately administered entities in those countries, have also come to differ in the qualifications required of their personnel, including age levels, and the requisite experience for entry or elevation into higher positions of responsibility. In most countries, research and extension groups are headquartered apart from each other. This separation, combined with the above factors, tends to foster separate conceptual approaches to dealing with the problems of farmers.

These two groups also tend to see each other very differently. Research scientists tend to view extension workers as unwilling to trust or accept research findings, unwilling to ask researchers for information, as not helpful in clarifying the nature and extent of field problems, and demanding of immediate answers to complex problems. Extension workers tend to think of research scientists, on the other hand, as secluded in their laboratories, unconcerned with the social and economic impacts of their research, noncommittal about their research results, and prone to obtuse language in their writings and reports.

Aside from how these two groups view each other, general studies have identified a number of weaknesses and deficiencies among research scientist and extension personnel. Researchers frequently do not specify which recommendations will maximize returns to various inputs or labor. They are frequently uninterested in the practical application of their findings. Most research scientists are more inclined to be concerned about publishing information that they deem useful to other scientists. This is probably attributable in part to the more centralized management system controlling research operations, which draws the attention of research scientists away from local definitions of needs and priorities.

Many research scientists are not prone to consider the long-term ecological consequences of their research, since their primary interest is in knowledge breakthroughs. In most of the less developed countries, agricultural research scientists tend to have an urban bias and to concentrate on international cash crops, with a commerical rather than a subsistence focus. In light of the continuing food production and distribution problems in such countries, it is easy to see why some question how many pure scientists the country can afford.

Extension workers similarly have been scrutinized and found lacking in practical farm experience or training appropriate to the task to which they have been assigned. Many are prone to working with the best farmers rather than with those who most need their services. Certain institutional factors in extension also hinder the work of extension personnel. Generally, the weak technical link with research means that the technical information being disseminated by extension staff may be out-of-date or no longer appropriate. Also, extension personnel tend to receive lower salaries than research scientists. Concurrent with that, extension staff usually carry an inferior status within the ministries or bureaus of agriculture and have less opportunity than research scientists to advance to positions of authority. Research scientists and extension staff see each other and have an opportunity to interact with each other during meetings, in joint field trials, during visits by extension staff to experiment stations, through farm visits, during short courses usually conducted by scientists, and during promotion or educational tours. The opportunities for interaction are generally there but various kinds of strains and tensions between the two groups prevent effective interaction.

Some of the tensions are mere human reflections of organizational strains. The two groups tend to differ with regard to basic professional definitions, focal concerns, and role perceptions. One group tends to emphasize the comprehension of phenomena while the other places more emphasis on changing situations. They tend to have different value considerations. Such communication factors as the lack of two-way direct relationships, information overloads, and differences in language or jargonistic styles also hinder their interaction. The two groups have different methodological grounding, conflicting views of the utility of practical experience and common sense, different standards for measuring achievement, and different views of the utility of negative criticism. Their orientation toward clients differs with regard to both understanding client needs and conceptualizing the nature of risks faced by clients. They also have and pursue different sources of recognition, have oftentimes perceived conflicts of interest, and have different loyalties and basic identities. Needless to say, the challenge of integrating the work of two such disparate groups is major.

The biggest problem may not be the fact that such strains and tensions exist, but rather the general failure to recognize them. The failure to see the need for more effective exchange between research scientists and extension personnel is of greater concern. The major goal of agricultural knowledge creation-diffusion-utilization systems or organizations is to help the producer and the consumer. Within this larger system, organizational members need to have common goals. But it is difficult to arrive at such common goals if such great personal differences exist. It has been said that the greater the differences in skills, prestige, and status among members of a work group, the more difficult it will be to establish and maintain satisfactory interpersonal relationships (Katz and Kahn 1966). A somewhat circular situation exists, then, wherein poor interpersonal relationships fuel the strain caused by differences in skills, prestige, and status, and these thwart efforts to arrive at common goals that might work to improve interpersonal relationships. An organizational factor, that is, the separation of research from extension, is more likely the major cause of these tensions than interpersonal differences per se.

Research scientists generally do not understand the role of extension agents. They fail to see that extension workers are supposed to be more than just technical experts, that they have more to extend than just scientific or technical information. Extension agents must be as responsible for social know-how as they are for technical know-how. Scientists fail to respect the fact that extension needs to be more responsive to clientele than do researchers or administrators. Extension workers are held accountable by clientele as sources of technical information. They are expected to provide farmers with information about seeds and the availability of various inputs. They are expected to provide such lab services as soil analysis and the analysis of plant and animal diseases. This diffuse set of expectations is exacerbated by the demand for result and method demonstrations and the system requirements of processing information regarding farmer problems and needs for the benefit of researchers and administrators. The role of the extension worker is, indeed, a very professional one, but seldom recognized as such by research scientists.

Extension workers also need to be helped to respect scientists' need for isolation and freedom. However, the tendency for agricultural scientists in Third World countries to become narrow is not a positive development. In the United States, agricultural scientists are sometimes forced to narrow their scope because of the publishing criteria related to tenure decisions. This is not functional for the Third World, however. The luxury of limited use of scarce resources is not there.

It may be that value gaps do exist between research and extension personnel, between scientists and practitioners. These gaps, although real, may be perceived by each group to be higher than they actually are. There are sufficient cases of effective informal exchange to suggest that extension and research personnel are able to overcome differences caused by membership in two different organizational entities. But contact on this individual level is not really sufficient to ensure that the farmer will receive good timely assistance. Organized and systematic interaction needs to occur, whether by involving extension staff in research planning as members of technical working groups, or by having research scientists participate in farmer training programs organized by extension.

Research and extension personnel tend to use different sources of information and to favor different kinds of communication. Scientists generally prefer journals, and extension personnel prefer face-to-face discussion. Somewhat because of this, researchers seldom, if ever, initiate contacts with extension workers. Extension workers are more prone to initiate contacts with scientists. It is sometimes useful to measure the extent and quality of research and extension interaction by assessing the reseacher's knowledge of farmer problems (conveyed to him by extension personnel) and the extension staff's knowledge of research activities (as might be observed during visits to experiment stations).

Regardless of how it is done, it is fundamentally important to agricultural development to achieve effective interaction between research and extension. In the People's Republic of China, research scientists have been forced to go to the village. This is perhaps effective if it is not done to such excess that it hinders the creative activity of the scientists. Such exposure to field situations has helped make Chinese research scientists sensitive to complex cropping systems (multiple, inter-, relay and vertical) as well as to the integrated pest management systems developed by peasant farmers (Stavis 1978).

The Philippines, as part of its Masagana 99 program, developed a "mini-kit" of packages of technology based on carefully conducted farm trials and supported by an array of services, inputs, and credit

(Athwal 1981).

Burma has evolved a "whole township" approach for integrating the functions of research and extension activities by establishing three-person township councils consisting of a technical representative (the extension agent), a political representative (member of the Burma Socialist Program Party), and a popular representative (a village leader)(U Kin Win et al., 1981). This approach suggests that authorities external to research and extension groups may have to force needed integration.

Korea's Office of Rural Development (ORD) has achieved an administrative integration of research and extension, as both groups are under its control. A further step was taken to define research and extension functions locally when research and extension work were linked to Korea's community organization program (Saemaul Undong). To achieve all this, a strong national commitment was needed.

From the foregoing discussion, we can see that the integration of agricultural research and extension functions in Third World countries presents an uneven picture, with some successes and some failures. The question remains as to how well we have learned those lessons and how well we are putting them to use today.

Present State of Affairs

Linking agricultural research to the needs and conditions of the rural majority continues to receive attention. During the past five or six years, more than forty developing countries have negotiated loans from the World Bank to finance the expansion and redirection of their agricultural extension services. These countries have adopted the Training and Visit (T&V) approach to extension (Benor 1970) sponsored by the World Bank. The T&V approach is designed to improve the level of agricultural production by large numbers of farmers who cultivate small farms. With this approach, village extension workers are provided regular fortnightly training by subject matter specialists and training officers and then proceed to visit a given number of contact farmers within the farming communities on a regular basis during the following two-week period. This activity is backed up by a series of research and extension committees operating at different administrative levels: national, regional, provincial. These committees develop recommendations for the extension service to transmit to farmers and guidelines for evaluating past experimental data and for designing a program of field trials to check out the results of experiments. Decentralized research and extension committees are also established at lower administrative levels in order to select recommendations that suit local conditions. Subject matter specialists play an important role on these committees. The membership of the decentralized research and extension committee is normally chaired by the district extension officer and includes subject matter specialists from the area, and staff from the nearest research station.

In addition to serving on such research committees, subject matter specialists are supposed to spend about one-third of their time visiting the nearest research substation and other research facilities. Their role is to keep informed of the latest developments in research and to bring the farmers' practical field problems to the attention of the research workers. They are also to collaborate with reseach workers, district extension officers, and village extension workers to design and carry out a program of field trials on farmers' fields.

Some recent performance evaluations of the T&V approach have shown that various governments are encountering considerable difficulty with its design. The current role of linking research and extension played by subject matter specialists (SMSs) is not being performed very well. The reality is that SMSs are diverted away from close and frequent interaction with researchers and farmers into more mundane and less significant administrative activities. A tremendous number of person-days per year are spent in the provincial office or traveling to and attending meetings and seminars. Little SMS time is spent in making individual forays to research stations or university or college campuses to seek specific answers to specific problems, nor to farmers' fields to conduct an in-depth analysis of any problems brought to their attention. Part of this may be explained by the small number of SMSs in each province (usually two instead of the recommended five), making sufficient specialization and geographic coverage impossible (Compton 1982).

Under the T&V approach, general advisory and/or technical research committees are supposed to exist at the provincial, district, and subdistrict levels. There is some evidence that agricultural problems are sometimes discussed in province and district-level meetings conducted by provincial administrators (governors and district chief administrative officers, respectively). Committees seldom exist that would focus specifically on the technical aspects of agriculture at these two levels. Seemingly little has been done to promote the development and use of committees at these different levels. Logically, the establishment of such committees should have been a prerequisite for the rest of the program; instead they appear to be a product of hindsight, and deemed a low priority one at that.

The need for such committees is a rather basic one, as there are prospects for increasing agricultural production by developing and providing suitable "packages" of services, recommendations, and inputs to fit the specific nature of different agroclimatic and agroeconomic zones. Concurrently, little effort is made by national-level administrators to encourage the adaptation of existing crop modules to varying soil and water conditions, district to district. To leave this adaptation to extension workers and local contact farmers is unrealistic. Subject matter specialists and researchers should be collaborating to do this, but it has generally been found that they are not.

In recent years considerable attention has been given to the extension-farmer linkage function within overall knowledge creation-diffusion-utilization systems (see Rogers chapter in this book). An equally serious need exists for a focus on research-extension linkage functions. The nature and extent of the research and extension linkage will ultimately affect the quality of the relationship between extension fieldworkers and farmers, themselves. One study of the agricultural research delivery system in Sierra Leone (Lakoh and Akinbode 1981) found that research and extension contacts are very limited and that village extension workers were largely dependent upon their immediate field supervisors for information about technological innovations. There was very limited contact between fieldworkers and research scientists because of the absence of established, formal links and a lack of general logistical support (transportation and per diem allowances to permit travel to experiment stations).

Given the urgency of the need to speed up information flows between research scientists, extension workers, and farmers, it appears that the present nature of agricultural research and extension programs in Third World countries represents a major obstacle to the achievement of development goals. There appears to be a general lack of understanding of the importance of differentiating various linkage roles within research and extension operations and finding effective means of coordinating the interaction of those roles. For example, there is a need to differentiate between the important role of the extension generalist and the extension subject matter specialist. Both roles are very important and cannot be performed by the same person.

The extension generalist must be a people-oriented person in contrast to the more technically knowledgeable and skilled specialist. The specialist serves as a technical adviser whereas the generalist also has to be concerned about his or her role as a social adviser. The generalist is usually located within a specific geographic area but specialists, largely because they are fewer in number, are forced to be mobile to operate within a larger area. The specialist represents a technical link to experiment stations and university-based technical agriculturalists. Generalists need to establish and maintain technical and functional links to universities and institutes where they can obtain the advice or assistance of social and behavioral scientists as well as technical scientists. The generalist comes to know and appreciate the indigenous knowledge of the farmers, themselves, while remaining somewhat aware of the wide range of institutionally organized, scientific knowledge available from a wide range of agricultural specialists. The specialist, being highly trained in a particular area, should ideally also be somewhat aware of the indigenous knowledge of farmers. Interaction between the generalist and specialist around these two bodies of knowledge represents a very important exchange.

The generalist must maintain a primary focus on human problems and a secondary concern for technical problems. The specialist, however, must maintain a primary focus on technical problems while being somewhat concerned about human problems. Needless to say, both specialist and generalist have a primary responsibility for facilitating the diffusion of indigenous and scientific knowledge, but the specialists' primary responsibility is to simplify and deliver scientific knowledge to the generalist. The generalist and the specialist need to rely upon each other in order to achieve a balanced response to problems with both technical and human dimensions. Generalists establish credibility in the eyes of farmers based on the extent to which they are perceived to be helpful to them. The credibility of the specialist is based more upon the achievement of technical results. It is the combination of roles, with the generalist having a primary allegiance to the people and the specialist having a primary allegiance to agency goals of increased production, that makes for a dynamic knowledge creation-diffusion-utilization system in agriculture.

Future Challenges and Opportunities

It is necessary to take into account the differences between U.S. experiences and present-day Third World necessities. Agricultural research in the United States was largely aimed at increasing production and commercializing the agricultural sector. The greater need of developing countries continues to be that of meeting the subsistence needs of their populations. Therefore, agricultural research has to be aimed at increasing production through an expansive diffusion of agricultural knowledge and information. We have seen how the science and practice of diffusion has progressed during the past several decades. We have moved away from the traditional focus on adoption behavior and related factors to a current emphasis on the ways in which innovations and the conditions for adoption are made available to potential adopters (Brown 1981). We need to go one step further to design ways to involve intended adopters or users of innovations in the very process of identifying and developing those innovations themselves. We need to see the farmer and not the researcher as the starting point for agricultural research. In the past we have tended to pursue a unilinear model of knowledge creation-diffusion-utilization in our thinking and action. Research activity has been seen as synonymous with the creation of knowledge, extension activity as being the same as diffusion, and farmer behavior as largely concerned with the adoption and utilization of innovations. This unilinear model needs to give way to more dynamic interactive models wherein farmers have a role in developing agricultural knowledge of relevance to their problems and situations.

We have been challenged to consider that knowledge exists only in the head of the knower. One person's knowledge is only information to another person and remains so until the other person internalizes and makes that information meaningful to his or her own life situation (Freire 1981). This definition of knowledge as "the act of knowing" represents an important challenge to research scientists and extension educators.

It is important that we begin to see farmers as professionals also (Chambers 1980). Recent scholarship has demonstrated the expansiveness and value of indigenous knowledge and learning systems (Brokensha et al. 1980). The Warren chapter in this book gives full treatment to this matter. It is important that research scientists and extension workers respect and capitalize upon the knowledge and experience of farmers themselves. In what are currently very important developments in farming systems research (FSR), it is also critical to include extension personnel in the design of those activities. Whereas research scientists are in scarce number in most Third World countries, extension staff blanket the countryside and are in daily contact with farmers. This provides them the opportunity to become familiar with farmers' knowledge and practices. Such knowledge needs to be processed by extension and fed back to research scientists, training officers, and administrators. Because of the nature and frequency of their contact with farmers, extension personnel are in a position to help research scientists identify critical farming problems and needs and to mobilize farmers into FSR activities as well. Farming systems research and related extension activities are not a replacement for more general extension work, but rather a means for bringing about the wide-scale linking of research, extension, and farming activities (see chapter by Waugh, Hildebrand, and Andrew in this book).

It is also important that we come to realize and accept the idea that, along with the existence of indigenous knowledge and learning systems, there are also people we might regard as indigenous specialists. Such new realizations or configurations present a series of challenges for extension. One of the most serious deficiencies in extension work today is the long and oftentimes discontinuous hierarchy between scientific knowledge and indigenous knowledge. Between the research scientist and the intelligent farmer are layers of extension personnel. Because of our general tendencies to evolve cognitive frameworks based on hierarchically oriented organizational charts, we generally fail to conceive of the existence and importance of lateral patterns, distributions, and networks that function at each level. This affects the functioning of knowledge systems in very serious ways. An awareness of the need for scientists to collaborate across disciplinary lines seems to be emerging (Russell et al. 1982). Interdisciplinary research methods are being identified and tested. However, such research is still only one half of the needed formula. The involvement of practitioners (and here I include farmers as practitioners along with extension staff) represents the other half of the formula—an organizational model we might call the "open natural system" (Scott 1981). Important advances have also been made in the theory and treatment of ecology as a new integrative discipline (Odum 1977). The importance of an ecological perspective on research, extension, and farming interactions is becoming increasingly clear. As we come to a clearer understanding of the rationale underlying the long-standing debate on the relative importance of basic research (for understanding) in contrast to applied research (for action), we are also becoming more aware of the existence and value of indigenous knowledge. We have begun to ask ourselves the questions: What can we learn from the farmers? What are some of the specific things researchers have learned from farmers? How can we learn from the farmer? What methods or techniques can we employ to learn what it is that the farmer knows? How can what we learn be made use of in agricultural development programs? In what ways can indigenous knowledge be integrated into the larger system?

From an extension programming perspective there are several promising strategies that might be employed to identify, recover, and make use of indigenous knowledge. The first is to identify and incorporate indigenous specialists into program planning and implementation in ways that make use of their talents. Concurrently, it may be possible to designate a subject matter specialist for the recovery of indigenous knowledge. Such as extension specialist could become the librarian for indigenous knowledge.

Another major and important strategy is the establishment of university-operated social laboratories. An effort would be made to identify, design, test, and promote alternative technologies and strategies for development. Research field operations and training functions would be conducted in a manageable geographic area and be closely related. Within the laboratory there would be an initial emphasis on relevancy, the basic idea being that it makes no sense to emphasize effectiveness in something that is irrelevant to local needs and conditions. Once it is possible to determine or ensure relevancy, effectiveness can be sought. As effectiveness is achieved, it then becomes possible to aim for efficiency in order to obtain a suitable cost-benefit ratio. Once such efficiency has been achieved or proven within a project or program area, it then makes sense to consider its expansion on a large-scale basis throughout a larger geographical area. Such a social laboratory can be a wise investment in the development of technical know-how, administrative know-how, and social and educational know-how. Once developed, such know-how can be built upon as time and resources permit.

Another practical way to improve the functioning of agricultural knowledge systems is to increase management capabilities. The question of making research and extension accountable to the farmer (Lundberg 1982) largely rests on the prospect for providing research managers who realize the need to disseminate the results of research. Scientists are generally just not prone to take an active role in this. In the past, research-extension liaison officers have been used in several countries. Oftentimes these officers had the responsibility to facilitate communication between research scientists and extension staff, but lacked the authority to require such interaction. There is a great need for managers of utilization systems, persons who are skilled in the management techniques necessary to facilitate the functioning of information and action systems. Perhaps we could go one step further and suggest that we need managers of research and diffusion systems. Sri Lanka's Regional Technical Working Groups (RTWG) are composed of representatives from research, extension, and the farming community and meet quarterly to deal with pressing problems and needs of the farming community. They are an ideal construct for achieving the kind of integrated functioning discussed in much of this chapter (Durfee 1980). The managerial function of the chairperson of the RTWG is a direct reflection of the role just described.

Another promising strategy is the field workshop employed by Cameron Clark in Bangladesh and other Asian countries. Here, adminstrators, researchers, educators, and farmers are brought together for field-based workshops to identify bottlenecks and promising solutions (Webster 1979). Systems evaluations need to be conducted in every province in order to identify program weaknesses and point out corrective measures. For example, relations and perceptions among staff and participants will differ at various levels (research scientist, program administrator, subject matter specialist, village extension worker, and farmer). This will, of course, adversely affect the efficiency of the system. The solution is to establish two "diagonal analysis teams" in each province, with the membership of the team rotated on a quarterly or semiannual basis. These teams are composed of persons who have no personal or otherwise direct interaction with each other, but who have played different roles in the same type of system. In the diagonal analysis team they work together to identify system weaknesses and propose remedies. Not knowing each other personally means that they will hesitate less in being honest and frank with each other as they analyze the functioning of the system. Perhaps an SMS will have to be appointed as team leader and instructed to ensure that social or official status distinctions are not allowed to override the free expression of candid views, i.e., that farmers and contact farmers are afforded as much opportunity to comment as district officers and research scientists. The team members chosen would assure a diagonal slice across a given province. By rotating the membership periodically (but keeping the same SMSs as team leaders), a picture of "system functioning" for the whole province could be obtained each year. This approach would complement decentralized monitoring and evaluation activities normally used for corrective action within each district.

There is much that can be done to improve the effectiveness of agricultural research and extension programs. But all the strategies for integrated rural development, decentralization, institutionalized research-extension linkages, on-farm research, and the accessing of indigenous knowledge systems, can only achieve a little in the face of landlessness and social stratification. Such conditions result in an unequal access to resources, a situation over which agricultural scientists and development workers can no longer assume a neutral stance. They must become proactive protagonists for change.

References

Arnon, Isaac. *Organization and Administration of Agricultural Research.* Amsterdam: Elsevier, 1968.

Athwal, D.S. "Strategy for Research and Extension in Agricultural Development." In *Accelerating Agricultural Development,* J.D. Drillon, Jr. and G.F. Saguiguit, eds. College, Laguna, Philippines: SEARCA, 1981, 98-112.

Axinn, George H. "Host Country Institutions and Diffusion of Technology." In *Transferring Food Production Technology to Developing Nations*, Joseph J. Molnar and Howard A. Clonts, eds. Boulder, Colo.: Westview Press, 1983, 146-53.

Benor, Daniel, "Generalists or Specialists? Extension Workers View the Problem." In *Agricultural Extension: A Sociological Appraisal*, Sara Molho and Marcia Gitlin, eds. Jerusalem: Keter, 1970, 32-50.

Benor, Daniel, and James Q. Harrison. *Agricultural Extension: The Training and Visit System.* Washington, D.C.: World Bank, 1977.

Beyer, Janice M., and Harrison M. Trice. "Current and Prospective Role for Linking Organizational Researchers and Users." Paper prepared for Conference on Producing Useful Knowledge for Organizations, University of Pittsburgh, October 28-30. Mimeo, 44 pp.

Bhatnagar, O.P. "Role Expectations, Role Performance, and Training Needs of Subject Matter Specialists Working in Jawahar Lal Nehru Agricultural University and State Department of Agriculture, Madhya Pradesh." Ph.D. dissertation, India Agricultural Research Institute, New Delhi, 1971.

Blase, Melvin G., and Arnold Paulsen. "The Agricultural Experiment Station: An Institutional Development Perspective." *Agricultural Science Review* 10 (Second Quarter 1972): 11-18.

Boyce, J.K., and R.E. Evenson. *National and International Agricultural and Extension Programs.* New York: Agricultural Development Council, 1975.

Brewer, J. "Flow of Communication, Expert Qualifications, and Organizational Authority Structures." *American Sociological Review* 36 (1971): 475-84.

Brokensha, David W., D. M. Warren, and Oswald Werner. *Indigenous Knowledge Systems and Development.* Lanham, Md.: University Press of America, 1980.

Brown, Emory J., and Albert Deekens. "Roles of the Extension Subject-Matter Specialist." *Rural Sociology* 23 (1958): 263-76.

Brown, Lawrence A. *Innovation Diffusion: A New Perspective.* New York: Methuen, 1981.

Chambers, Robert. "Two Frontiers in Rural Management: Agricultural Extension and Managing the Exploitation of Communal Natural Resources." IDS Communication 113. Brighton, England: University of Sussex, 1975.

——. "Understanding Professionals: Small Farmers and Scientists." IADS Occasional Paper. Washington, D.C.: International Agricultural Development Service, 1980.

Compton, J. Lin. "Evaluating the Training and Visit System: a Recent Experience." In "Evaluation Report on Extension Outreach Project" prepared by the Agricultural Office, U.S. Agency for International Development, Bangkok, Thailand, 1982.

——. *Report of Consultant on Extension and Communications to Farming Systems Development Project - Eastern Visayas (Philippines).* Baybay, Leyte: Visayas-State College of Agriculture, July 1983.

——. "Linking Scientist and Farmer: Re-Thinking Extension's Role." In *World Food Issues*, 2nd edition, M. Drosdoff, ed. Ithaca: Cornell University Program in International Agriculture, 1984a, 79-84.

——. "Extension Program Development." In *Agricultural Extension: A Reference Manual,* B. Swanson, ed. Rome: FAO, 1984b, 108-19.

Coward, E. Walter. "Irrigation Development: Technology, Traditions, and

Transfers." In *Transferring Food Production Technology to Developing Nations*, Joseph J. Molnar and Howard A. Clonts, eds. Boulder, Colo.: Westview Press, 1983, 154-66.

Cummings, Ralph W., Jr. "Strengthening Linkages Between Agricultural Research and Farmers: An Overview." A paper prepared for the Organization for Economic Cooperation and Development's (OECD) Workshop on Linkages Between Agricultural Research and Farmers in Developing Countries. Paris, April 1981.

Dubey, V.K. "A Study of the Integrated Functioning of Teaching, Research, and Extension in Selected Agricultural Universities." Ph.D. dissertation, India Agricultural Research Institute, New Delhi, 1972.

Duncan, W. Jack. "The Researcher and the Manager: A Comparative View of the Need for Understanding." *Management Science* 20 (1974): 1157-63.

Dunn, William N. "The Two-Communities Metaphor and Models of Knowledge Use." *Knowledge: Creation, Diffusion, Utilization* 1 (1980): 515-36.

Durfee, Art. "Re-organization of Research, Extension and Training Resources for Agriculture." Report of extension consultant to the Government of Sri Lanka, Colombo, Sri Lanka, 1980.

Elkana, Yoseph. "Research-Extension Links in Agriculture: A Study of the Communication Behavior of Professionals." In *Agricultural Extension: A Sociological Appraisal*, Sara Molho and Marcia Gitlin, eds. Jerusalem: Keter, 1970, 51-76.

Evenson, R.E., and Y. Kislev. "Investment in Agricultural Research and Extension: A Survey of International Data." *Economic Development and Cultural Change* 23 (1975): 507-21

Fairweather, George. "The Prospective Center for Experiments and Training." In *Methods for Experimental Social Innovation*. New York: Wiley, 1967, 213-28.

FAO (Food and Agriculture Organization). *1977 FAO Production Yearbook*.

——. *China: The Agricultural Training System*. Rome: 1980.

Fishel, Walter L., *Resource Allocation in Agricultural Research*. Minneapolis: University of Minnesota Press, 1971

Flinn, William L., and Frederick H. Buttel. "Sociocultural Constraints on the Transfer and Adoption of Agricultural Technologies in Low Income Countries." In *Transferring Food Production Technology to Developing Nations*, Joseph J. Molnar and Howard A. Clonts, eds. Boulder, Colo.: Westview Press, 1983, 125-45.

Freire, Paulo. "Extension or Communication." *Education for Critical Consciousness*. Part 2. New York: Continuum, 1981, 91-164.

Friedland, William H., and Tim Kappel. *Production or Perish: Changing the Inequities of Agricultural Research Priorities*. Santa Cruz: University of California, 1979.

Gabelman, Warren H., and R.R. Romanowski. "Research and Extension." In *Vegetable Farming Systems in China*, Donald L. Plucknett and Halsey L. Beemer, Jr., eds. Boulder, Colo.: Westview Press, 1981, 171-83.

Gidley, V.N. "Towards Better Liaison: A Survey of Communication Between Agricultural Research and Extension." *Journal of the Australian Institute of Agricultural Science* 43 (September-December. 1977): 110-16.

Gliessman, Stephen R. "Research Management in Traditional Tropical Agroecosystems in Southeast Mexico." In *Agricultural Sustainability in a Changing World Order*, Gordon K. Douglass, ed. Boulder, Colo.:

Westview Press, 1983.

Hargreaves, Lesley E. "Communication Between Agricultural Scientists and Extension Workers: A Study in the Sheep and Wool Branch of the New South Wales Department of Agriculture." *Agricultural Extension Bulletin* No. 1. Department of Agricultural Economics, University of Sydney, 1976.

Hart, Robert D. "An Ecological Systems Conceptual Framework for Agricultural Research and Development." In *Readings in Farming Systems Research and Development*, W.W. Shaner, P.F. Philipp, and W.R. Schmehl, eds. Boulder, Colo.: Westview Press, 1982.

Havelock, Ronald G. *Planning for Innovation: A Comparative Study of the Literature on the Dissemination and Utilization of Scientific Knowledge.* Ann Arbor: Center for Research on Utilization of Scientific Knowledge, Institute for Social Research, 1969.

Hildreth, R.J. "Tensions Between Research and Extension Workers: Three Hypotheses." *Journal of Farm Economics* 47 (1965): 838-40.

Honadle, G.H., S. Tip Walker, and Jerry M. Silverman. "Institutional and Organizational Constraints." In *Implementing Rural Development Projects*, Elliott R. Morss and David G. Gow, eds. Boulder, Colo.: Westview Press, 1983.

Jaiswal, N.K., and K.B. Srivastava. "Transfer of Farm Technology to Small Farmers: A Study of Organizational Problems in India." *Agricultural Administration* 3 (October 1976): 251.

Johnston, B.F., and P. Kilby. *Agriculture and Structural Transformation: Economic Strategies in Late-Developing Countries.* New York: Oxford University Press, 1975.

Katz, Daniel, and Robert L. Kahn. *The Social Psychology of Organizations.* New York: Wiley, 1966.

Kidd, D.W. "A Systems Approach to Analysis of the Agricultural Extension Service of Western Nigeria." Ph.D. dissertation, University of Wisconsin, 1971.

Lakoh, Alpha K., and I.A. Akinbode. "Agricultural Research Delivery System in Sierra Leone." *Agricultural Administration* 8 (1981): 289-305.

Leonard, D.K., and D.R. Marshall, eds. *Decentralization and Linkages in Rural Development.* Berkeley: Institute of International Studies, University of California, 1983.

Lundberg, A. David. "Smallholder Agricultural Development: The Role of Research and Extension." M.P.S. paper, Cornell University, 1982.

Mann, Floyd, and Rensis Likert. "The Need for Research on the Communication of Research Results." *Human Organization* II (1952): 15-20.

Mejia, Ruby Mariela. "An Historical Analysis of Colombia's Agricultural Extension Programs and Their Underlying Assumptions: 1950-1980." M.S. thesis, Cornell University, 1983.

Mitchell, John J. "Problems and Solutions of Indigenous Knowledge Recovery." M.P.S paper, Cornell University, 1982.

Morss, Elliott R., John K. Hatch, Donald L. Mickelwait, and Charles F. Sweet. *Strategies for Small Farmer Development: An Empirical Study of Rural Development Projects in The Gambia, Ghana, Kenya, Lesotho, Nigeria, Bolivia, Colombia, Mexico, Paraguay and Peru.* Boulder, Colo.: Westview Press, 1976.

Moseman, A.H., ed. *National Agricultural Research Systems in Asia.* New

York: Agricultural Development Council, 1970.

Nagel, Uwe Jens. *Institutionalization of Knowledge Flows: An Analysis of the Extension Role of Two Agricultural Universities in India.* Frankfurt: DLG-Verlag, 1980.

Netting, Robert. M. "Agrarian Ecology." *Annual Review of Anthropology* 3 (1977): 21-56.

Odum, E.P. "The Emergence of Ecology as a New Integrative Discipline." Science 195 (1977): 1289-93.

Oram, Peter. "New Opportunities for Blending Science and Tradition in Agriculture." In *The World Food Problem and Cooperation in Agricultural Research.* Wageningen: International Agricultural Center,

——. "Collaboration Between National and International Institutions in the Development of Improved Agricultural Technology." In *Science, Politics, and the Agricultural Revolution in Asia,* Robert S. Anderson, et al., eds. Boulder, Colo.: Westview Press, 1982, 379-425.

Parker, J.S. "Communication Between Research and Extension Workers: Agricultural Research." *New Zealand Agricultural Science* 5 (November 1970): 10-12.

Pray, Carl E. "The Green Revolution as a Case Study in Transfer of Technology." In *Technology Transfer: New Issues, New Analysis,* Alan W. Heston and Howard Pack, eds. Special issue of *The Annals of the American Academy of Political Science,* Vol. 458 (November 1981): 68-80.

Rice, E.B. *Extension in the Andes: An Evaluation of Official U.S. Assistance to Agricultural Extension Services in Central and South America.* Cambridge: MIT Press, 1974.

Rogers, Everett M. and F. Floyd Shoemaker. *Communication of Innovations: A Cross-Cultural Approach,* 2nd ed. New York: Free Press, 1971.

Rogers, Everett M., and Jane E. Marcus. "Advances in Diffusion Theory." In *Knowledge Utilization Systems in Education,* William J. Paisley and Matilda Butler, eds., Beverly Hills: Sage, 1983, 251-57.

Roling, Niels. "Knowledge Brokerage for Increasing the Relevance of Agricultural Research for African Smallholders." *Rural Africana* 27 (Spring 1975): 113-19.

Rothman, Jack. *Planning and Organizing for Social Change: Action Principles from Social Science Research.* New York: Columbia University Press, 1974.

——. *Using Research in Organizations: A Guide to Successful Application.* Beverly Hills: Sage, 1980.

Russell, Martha Garrett, with Richard J. Sauer and John M. Barnes, contributing eds. *Enabling Interdisciplinary Research: Perspectives from Agriculture, Forestry and Home Economics.* University of Minnesota, Agricultural Experiment Station Publication 19, 1982.

Ruttan, Vernon. "Organizing Research Institutions to Induce Change: The Irrelevance of the Land Grant Experience for Developing Economies." Institute of Agriculture, Journal Paper Series, No. 1313. St. Paul: University of Minnesota, 1968.

——. "Technology Transfer, Institutional Transfer, and Induced Technical and Institutional Change in Agricultural Development." In *Agriculture in Development Theory,* Lloyd G. Reynolds, ed., New Haven: Yale University Press, 1975, 165-91.

——. "Induced Innovation and Agricultural Development." In *Agricultural*

Sustainability in a Changing World Order, Gordon K. Douglass, ed. Boulder, Colo.: Westview Press, 1983.

Schultz, T.W. *Transforming Traditional Agriculture.* New Haven: Yale University Press, 1964.

Scott, W. Richard. *Organizations: Rational, Natural, and Open Systems.* Englewood Cliffs, N.J.: Prentice-Hall, 1981.

Sharma, J.C., and A.W. Sohoni. "District-Level Subject-Matter Extension Specialists' Job in the Farmers' Advisory Service." Hissar, India: Haryana Agricultural University, 1971.

Stavis, Benedict R. "Agricultural Research and Extension Services in China." *World Development* 6 (May 1978): 631-45.

——. *Turning Point in China's Agricultural Policy.* Michigan State University Agricultural Economics Rural Development Series, Working Paper 1, 1979.

Steinberg, David I., et. al. *Korean Agricultural Research: The Integration of Research and Extension.* USAID Project Impact Evaluation No. 27. Washington, D.C.: USAID, 1982.

Sundquist, James L. "Research Brokerage: The Weak Link." In *Knowledge and Policy: The Uncertain Connection*, Laurence E. Lynn, Jr., ed. Washington, D.C.: National Academy of the Sciences, 1978, 126-44.

Swaminathan, M.S. "Role of National Programs in Linking Research and Development." In *Proceedings of the International Symposim on Development and Transfer of Technology for Rainfed Agriculture and the SAT Farmer*, Vrinda Kumble, ed. Patancheru, India: ICRISAT Center, 1979.

Tang, A. M. "Research and Education in Japanese Agricultural Development, 1880-1938." *Economics Studies Quarterly* 13 (February-May 1963): 27-41 and 91-99.

Trigo, E.J., M.E. Pineiro, and J. Sabato. "Technology as a Social Issue: Agricultural Research Organization in Latin America." In *Technical Change and Social Conflict in Agriculture: Latin American Perspectives*, M. E. Pineiro and Ed. J. Trigo, eds. Boulder, Colo.: Westview Press, 1983.

Tushman, Michael L., and Thomas J. Scanlan. "Boundary-spanning Individuals: Their Role in Information Transfer and Their Antecedents." *Academy of Mangement Journal* 24 (1981): 289-305.

U Khin Win, U Nyi Nyi, and E.C. Price. "The Impact of a Special High-Yielding-Rice Program in Burma." IRRI Research Paper Series No. 58, Los Banos, Laguna, Philippines, 1981.

Webster, B.N. "Linkages for Transfer of Technology." In *Proceedings of the International Symposium on Development and Transfer of Technology for Rainfed Agriculture and the SAT Farmer*, Vrinda Kumble, ed. Patancheru, India: ICRISAT Center, 1979.

Winkelmann, Donald, and Edgardo Moscardi. "Aiming Agricultural Research at the Needs of Farmers." In *Readings in Farming Systems Research and Development*, W.W. Shaner, P.F. Philipp, and W.R. Schmehl, eds. Boulder, Colo.: Westview Press, 1982.

Wolf, W.C., Jr. "Linking Knowledge Producers and Knowledge Users." Document No. ED 151 342. Alexandria, Va.: Educational Document Reproduction Service, 1978.

Wyckoff, J.B. "Closer Cooperation Between Research and Extension." *Journal of Farm Economics* 47 (1965): 834-37.

CHAPTER 6

Evolution and Transfer of the U.S. Extension Model

EVERETT M. ROGERS

At the heart of the research/extension/farmer communication process is the interaction between a local extension agent and a farmer. For this interpersonal communication to be effective, there must be (1) useful information for the extensionist to convey to the farmer (usually this information comes from agricultural research), and (2) an adequate number of local extension workers so that they can personally communicate with many farmers. Unfortunately, these two conditions are not always met in industrialized nations like the United States, and they are even less likely to be found in Third World nations, where extension/farmer ratios are often 1 to 10,000 or more (Rice 1974). How would you like to contact 10,000 farmers?

The Agricultural Extension Model

The U.S. agricultural extension model is one of the most widely recognized systems for the diffusion of innovations in the world today. Probably no other government or private agency can claim to be more successful in transferring technology. The extension services are credited with being spectacularly successful in diffusing agricultural research results to farmers, and thus raising their levels of productivity. Since the 1950s, the U.S. extension model has been adopted in many Third World nations. Later we shall question whether this reputation for success is completely deserved.

The agricultural extension model is a set of assumptions, principles, and organizational structures for distributing the results of agricultural research to farm audiences in the United States. This model is based directly on the experience of the U.S. government agency responsible for diffusing agricultural innovations and closely parallels the conventional conception of a research and development/diffusion/utilization process.

Historical Development of the Cooperative Extension Service

Both the U.S. Department of Agriculture (USDA) and land-grant universities were begun by federal legislation passed in 1862. What is now the USDA had previously existed as a small part of the U.S. Patent Office. Several attempts to pass a land-grant college bill had been made previous to 1862, but these laws were consistently defeated by states' rights legislators who viewed the federal grant of land to a state as a threat to the state's power. During the Civil War, when the southern states' rights enthusiasts were absent from Congress, the Morrill Act was quietly passed. This law provided that in each state an acreage of federal land was to be set aside, and the income from this property would support a state college or university for teaching "agriculture and the mechanic arts."

Actually, the Morrill Act was a general, vague, and ambiguously worded document, and its meaning was not clear even to Senator Morrill's fellow senators. Historians suggest that this may have been deliberate. To gain the political support that would ensure its passage by Congress, the act may have been purposely designed to be vague. This lack of precision and clarity in the original land-grant college legislation has made it possible for these colleges to adjust and adapt to changing socioeconomic conditions.

The early professors of scientific agriculture in the land-grant colleges soon realized that they lacked sufficient knowledge about many agricultural topics. This realization caused a demand for agricultural research on farm problems in agronomy, animal breeding and nutrition, horticulture, and other fields. The Hatch Act of 1887 provided federal funds to state agricultural experiment stations, and these research centers developed in most states as an adjunct to the land-grant college of agriculture. The director of the experiment station is usually responsible to the dean of agriculture, the university president, and the state government. The USDA annually reviews the research projects in each state to ensure that they meet federal stipulations. Today an average of about one dollar out of four spent at the state experiment stations is of federal origin.

Need for an Agricultural Extension Service

In the early 1900s the state extension services were established as a further addition to the land-grant colleges and universities. Starting in 1914, the Smith-Lever Act provided federal grant-in-aid funds to each station for extension purposes. The extension workers in each state are staff members of the land-grant university. The state extension services work closely with the college of agriculture teaching personnel and the experiment station research workers. In about half the states, the dean of agriculture at the land-grant college is responsible for classroom teaching, the experiment station, and the extension service. In other states, there may be a separate director of the extension service, but this individual usually works closely with the dean of agriculture. In essence, the agricultural extension service in the United States is attached as an outreach activity to each land-grant university's college of agriculture.

The professors of agriculture in the land-grant colleges wanted their research-based knowledge to be used by farmers. Until about 1914, there was no effective means of transmitting such information from the colleges at Ames, Lansing, Davis, Raleigh, Ithaca, and elsewhere, to farm audiences. The agricultural professors wrote bulletins aimed at farmers, many of whom could not read or understand them; and there was no effective system for their distribution. Special campaigns were organized around a trainload of college professors who would tour their state to stump local communities in favor of some agricultural innovation, like seed corn selection in Iowa. Their knowledge was often received eagerly, but follow-up was lacking, and eventually the "corn seed gospel train" approach was dropped. Lastly, agricultural improvement societies were formed in local communities to promote agricultural innovations among the gentleman-farmer class. Their main forums were county and state fairs, which provided rural entertainment but much less effective agricultural diffusion. During this period of experimentation, a variety of piecemeal approaches to effective diffusion were tried, and all found wanting. The agricultural extension model was not yet on the scene.

Credit for the first extension approach of a lasting nature, and the format still largely followed today, is accorded Broome County, New York. The Binghamton Chamber of Commerce in Broome County was concerned about the welfare of local farmers, as agriculture was the main local industry. In 1911 the Chamber's "Farm Bureau" (so named in accordance with such other divisions of the Chamber of Commerce as the Roads and Alleys Bureau, the Protection Bureau, etc.) decided to employ a recent agricultural graduate from Cornell University to diffuse innovations to farmers in the county. Part of his salary the first year was donated by the local Delaware and Lackawanna Railroad, and so he was called a "county agent," as all railroad employees were called "agents." The Binghamton Chamber of Commerce's Farm Bureau included several leading farmers who solicited their neighbors for donations to help pay the county agent's yearly salary, and soon these donations were institutionalized into annual memberships in the Farm Bureau.

Note that even at this early date the agricultural extension service expressed

1. The felt need for a more rapid diffusion of agricultural innovations
2. Strong local control over the professional change agent by an organization of farm leaders
3. Responsiveness to local elites among the client system

The idea of a county agricultural agent, and of a local farm bureau, spread rapidly across the United States after 1911. The movement was spurred (1) by provision of federal financing through the Smith-Lever Act of 1914, and (2) by the need for higher agricultural production during World War I.

The early development of extension services is closely related historically to the development of farm bureaus, from formation of the first county farm bureau for educational purposes in 1911. The parallel development continued through the growth of the county agent movement during World War I. As the county agent movement spread, so did the farm bureaus as their sponsoring body. The turning point in extension service-farm bureau relationships was reached when the county farm bureaus federated into state organizations and then into a national organization. After the formation in 1919 of the American Farm Bureau Federation (AFBF), relationships were severed in most states (but not until 1954 in Iowa and Illinois). The new organization operated as a legislative pressure group on the state and national level, a function incompatible with its original purpose of local sponsorship for county extension agents. Although division of the extension services from the farm bureaus began in 1919, they have remained friendly to this day, and legislative assistance from the AFBF and state farm bureaus is one reason for the financial strength and stability of the state and federal extension services.

The growth of funding and personnel invested in the agricultural extension effort since 1920 has been striking. Although federal financing of extension did not begin until the Smith-Lever Act of 1914, as we have noted, by 1920 there were already over 3,000 employees in the extension services. Today the total is about 17,000 full-time equivalents, with about 12,000 county and area extension agents, 4,000 state-level extension specialists, and about 1,000 administrators.

Funds for the extension services come from federal, state, and local sources. The growth of the federal share has been almost exponential, a rate of growth generally characteristic of federal spending. Part of the increase in this funding can be attributed to inflation. Although the current federal extension appropriation is more than four times that of 1950, the real program increase is only about twice the 1950 level. It is a measure of the popularity of the Cooperative Extension Service that it has managed to maintain a high level of growth at a time when federal spending on agriculture generally has not kept pace.

The federal investment in agriculture extension represents a heavy commitment compared to spending in agricultural research. About 40 to 60 percent of the annual federal investment in agricultural research and development (R&D) goes to federally funded extension activities. For example, the USDA recently allotted $423 million for research and development, and the total would reach over $600 million if state funding were also included. The annual federal budget for extension was $200 million; with state and county government contributions, the total annual budget for the extension services was about $500 million. Thus the total extension budget almost approaches the total public agricultural R&D budget. Even if only the federal investment is considered, extension receives about half the funding of R&D. In other fields the comparable figures for federal extension-type activities as a proportion of federally supported R&D are much smaller.

- Law Enforcement Assistance Administration: 14%
- National Institute of Education: 10%
- U.S. Department of Labor: 3%
- National Institute of Mental Health: 2%
- National Aeronautics and Space Administration (NASA): 0.17%

Undoubtedly one of the reasons for the success of the agricultural extension services is their relatively high, stable budget. The financial success is, in turn, aided by the support given to the agricultural extension services by the powerful American Farm Bureau Federation.

Local Control and Needs Orientation

At the county level, county extension agents needed a local body to replace the farm bureau as their organizational tool for conducting educational activities. A county advisory board, composed of fifteen to twenty-five local leaders, was formed to meet with the county extension staff on a monthly basis. The extension service has such a local policy making group in most counties. This group, with the county extension agents, decides the nature of the extension program each year. It also helps develop the budget for the extension service in the county, and may have some influence in the hiring or firing of the county agents.

County extension advisory groups provide client participation in the planning of extension service activities. For instance, if local farmers have a concern about the ecological problems caused by insecticides and fertilizers, they will see that this topic is a county extension activity in the year ahead. The county extension agents, dependent on client interest for the success of their diffusion efforts, are naturally inclined to be attuned to farmers' needs.

The county extension advisory groups can also initiate a bottom-up request for agricultural research. For instance, perhaps they find that there is little scientific information available about pesticides. The county agent will convey a request for information to the state agricultural university, and if it cannot be answered with existing research-based knowledge, the state agricultural experiment station may initiate a research project on this topic. While such "extension-in-reverse" certainly occurs, the actual cases of such needs leading to new research projects are probably few.

Rise of the State Extension Specialists

Of the 17,000 professional staff members at all levels in the U.S. extension services, only about 12,000 are local-level extension agents. Practically all of the 3,150 counties in the United States have an agricultural agent (who is usually also the county extension director, or administrative chief of the county-level employees). About 80 percent of the counties also have a home economics agent, and a considerable number have a 4-H Club agent, and/or marketing agent.

What do the other 5,000 professionals who are not county-level agents do? Some 1,000 or so are administrators or supervisors at the district, state, or federal level. The remaining 4,000 or so are extension subject matter specialists at the state level. These specialists attempt to interpret current research findings in their particular field to the county extension employees, and thus eventually to the extension clientele. The development of the specialists began slowly. Before 1955 there were never more than 2,000 specialists, but since then there has been a considerable expansion of this specialist corps. Over two-thirds of the specialists are still in agricultural fields, but much of the expansion in recent years has come in nonagricultural fields. The rapid growth of the specialists coincides with the diversification of the extension service program and its audiences.

How do extension specialists relate to the rest of the extension system? Consider extension agronomy specialists. Their offices are typically in the department of agronomy at the state agricultural college, and hence they are colleagues of the research and teaching faculty there. They travel throughout the state to address farmer meetings and keep county extension agents abreast of new developments in agronomy. Likewise, there are extension specialists in farm management, marketing, rural sociology, animal husbandry, poultry husbandry, home economics, and other fields. The number and importance of extension specialists increased as the nature of farming became more specialized.

Essentially, specialists transmit research-based knowledge to the county extension agents. They are the county agents' agents. To effectively fill this function, the specialists must be able to bridge the scientific/intellectual system of the state agricultural university with the pragmatic world of the farmers and their county agents. Usually the specialist has had previous experience as a county extension agent, and then moved, through graduate study, into a specialized agricultural field. About half of the extension specialists have Ph.D. degrees, and many of the rest are doctoral candidates on a part-time basis. This high level of scientific training is necessary for the specialist to be able to understand research publications and to decode their contents into language that county agents and their farmer-clients can understand and apply.

The success of the extension specialists rests directly on how well they can serve the needs of their clients. Similarly, the reward system for research workers in the USDA and the state agricultural experiment stations strongly encourages finding research results and publishing them in a form useful to farmers. This pressure toward practical research facilitates the work of state extension specialists who must cooperate with research and teaching colleagues in their academic departments. All are pulling together to produce useable knowledge, and get it diffused to and adopted by farmers. Were agricultural research activities not oriented toward potential use, the linking function of the extension specialist (and the county agent) would be much more difficult.

In conclusion: the specialization of U.S. agriculture helped create a need for extension specialists; the extension model was hence adjusted to cope with an environmental change in the nature of agriculture.

Broadening the Subject Matter of Extension

In 1911 the sole content of extension activities was biological agriculture: agronomy, animal husbandry, poultry. County agents placed their main emphasis on production technology like new seeds, fertilizers, improved livestock, and farm equipment.

Soon the farmer's wife began to demand information on nutrition, child care, and home management. Accordingly, the county agricultural agent—usually with a college degree in technical agriculture, was paired with a county-level colleague—the home economics agent—usually with a bachelor's degree in home economics from a land-grant college. Home economics staff now make up about one-quarter of all extension personnel in the United States.

County extension staffs were later expanded to include county agents responsible for 4-H Club activities. These 4-H Club agents usually possessed a bachelor's degree in technical agriculture, and were considered future county agricultural agents in training. After five to seven years as 4-H agents, they might be promoted to agricultural agents.

By the 1930s and 1940s, many agricultural leaders began to see that agricultural production technology was only a part of the total picture. What good were production-increasing innovations when U.S. agriculture was faced with a farm surplus? State extension services began to turn some of their attention to problems of agricultural marketing and to consumer information programs.

These changes marked a trend away from a focus on strictly agricultural production to the addition of social science subject matter related to agriculture. The shift away from sole dependence upon biological (production) agriculture also meant that the extension service appealed to new, nonfarm audiences. Since the Smith-Lever Act of 1914 (federal aid to the station extension services) did not specify that the extension service should work only with farm people, it left room for an increasing trend to work with urban and rural nonfarm audiences in recent years.

Rise of the Extension Aides

Until 1969 all extension agents were professionals, holding (at least) university degrees in agriculture or home economics. It was assumed that technical competence based on professional training was fundamental for earning a high level of credibility in clients' eyes.

The federal War on Poverty programs of the 1960s originally concentrated on urban core areas, although about a third of all U.S. poor lived in rural areas. By the late 1960s, there was much pressure on extension services to use their educational efforts to solve rural (and urban) poverty problems. The availability of additional federal funds to the extension services for this purpose led to the rise of extension aides.

The aide approach was launched with the Expanded Food and Nutrition Education Program (EFNEP) in 1969. This program has reached more than one million families, about two-thirds of whom are from minority groups. About 7,000 aides are employed. The aides are an adjunct to the county home economics agent and provide home visits to low-income housewives on a regular and intensive basis. The aides are themselves low-income—selected from among the client audience. They are given brief in-service training and close supervision and are required to make a regular schedule of home visits. The original objective of the program was to improve the nutritional status of poor families, and the aides were officially called "extension nutrition aides." In practice, the problems of their clients could not be so easily categorized and the aides have sought to provide information on whatever needs they encountered: family planning, how to become eligible for welfare payments, child discipline, budget management, and so on.

Typically, fifteen to twenty aides, directly led by two or three aide-supervisors, might work under the direction of a county home economics agent. The aides are drawn from the local community, and unlike many extension professionals, they are often members of the minority groups that they serve. Many work in urban areas, where home visiting does not require automobile transportation.

The aide approach, widely used in other federally sponsored war on poverty programs with the urban poor, recognized that in addition to credibility based on professional training, trustworthiness could derive from a high degree of socioeconomic similarity with clients. The aides could "talk the clients' language" in a manner that the home economics extension agent seldom could.

The Expanded Food and Nutrition Education Program was undoubtedly one of the most radical single changes in the Cooperative Extension Service in the past fifty years. Some extension professionals originally opposed EFNEP, and some still do today. It was a radical change for the Cooperative Extension Service in almost every respect:

1. EFNEP had almost nothing to do with production agriculture; previously food and nutrition had been given only limited attention by home economists in the extension service.
2. The use of indigenous aides was a "first" in the Cooperative Extension Service. This aide strategy put the professional home economist in the role of supervisor, a new responsibility.
3. The targeting of rural and urban lower-income families, with the attendant goals of equity, was a sharp departure from extension's previous focus on prosperous farm families, with a goal of increased farm production.
4. The educational method of home visits on an intensive basis with 50 to 150 clients per aide contrasted with the extension service's prior tendency toward more mass educational methods like group meetings.

Criticism of Extension

Despite the general image of success usually attached to the agricultural extension model in the United States, it has had detractors and critics since its origin. Some criticisms were directed at its elite bias, its continued emphasis upon agricultural production in the face of farm surpluses and a decreasing farm population in the United States, and its close friendship with the American Farm Bureau Federation.

But certainly the strongest critique was published in 1973 under the ingenious title of *Hard Tomatoes, Hard Times: The Failure of America's Land Grant College Complex*. The author, Jim Hightower, headed a Washington-based organization called the Agribusiness Accountability Project. The book's title derives from a specific research project at one state agricultural experiment station that bred new tomato varieties to facilitate machine-picking, an innovation of particular advantage to large-scale tomato farmers. The book's theme is that much of the tax-supported agricultural research at state agricultural universities is designed to serve the needs of agribusiness corporations and the largest commercial farms. The state agricultural university-extension service complex is a closed system, Hightower claims, whose priorities are set through interlocking relationships of elite farmers and agribusiness firms, with little voice for the interest of the environmentalist, consumer, family farmer, or rural poor.

Hightower's argument raised important questions about the traditional criteria for evaluating the success of the agricultural extension model. Undoubtedly the extension services have played an important part in increasing the productivity of U.S. agriculture, especially since about 1940. But the extension services have not greatly assisted rural people in coping with the consequences of the agricultural revolution that the services helped unleash.

Perhaps in an aggregate sense, the extension services in the United States have only helped farmers as a sector, if we judge the rate at which farmers have passed out of farming. At the same time that the extension service contributed to the technological push that phased so many farmers out of agriculture, other federal agencies labored to help young farmers enter the occupation (through vocational agriculture programs that provide high school training to future farmers, and through the Farmers Home Administration, which gives credit and management advice to young and low-income farmers), and to stay in it (through the Agriculture Stabilization and Conservation Service, which seeks to control farm surpluses and thus buoy up agricultural prices). All of these agricultural agencies are headquartered in the U.S. Department of Agriculture, where they compete for funds. Critics of the extension service point out that while it has sought to raise average levels of farm income, it has done so in a way that redistributed incomes toward greater inequality. Other federal agricultural agencies have just the opposite priority: to increase the equality of farm income distribution.

Lessons Learned from the Domestic Experience

In this review of the development of the Cooperative Extension Service, we have noted a number of different stages in the development of the system and a number of elements which appear at different stages in different forms. To summarize our review, we present here an outline of these elements and how they have changed over time.

We believe there are eight main elements in the U.S. agricultural extension model.

1. A critical mass of new technology, so that the diffusion system has a body of innovations with potential usefulness to practitioners.
2. A research subsystem oriented to application, as a result of the incentives and rewards for researchers, research funding policies, and the personal ideologies of the researchers.

3. A high degree of user control over the research application process, as evidenced through client participation in policy determination, attention to user needs in guiding research and extension decisions, and the importance accorded feedback from clients on the system's effectiveness.

4. Structural links among the technology transfer system's components, as provided by a shared conception of the system, use of the common "language" by members of the system, and by a common sense of mission.

5. A high degree of client contact by the linking subsystem, which is facilitated by reasonable agent-to-client ratios and by a relatively homogenous client audience.

6. A "spannable" social distance across each interface between components in the system. Social distance is reflected in professionalism, formal education, technical expertise, and specialization. Generally, those variables decrease in degree as one moves from the research subsystem (where Ph.D.s are usually employed), through linkers, to the client subsystem of farmers and rural people.

7. Evolution of extension as a complete system for technology transfer, rather than as a research utilization system grafted on as an additional component of an existing research system.

8. A high degree of control by the system over its environment, enabling the system to shape the environment rather than passively reacting to changes in this environment. Such technology transfer is less likely to face unexpected crises or competitors, and can obtain adequate resources. The degree of control is expressed through the system's power base, its perceived legitimacy, and its political-legal influence.

What generalizations may be offered about the agricultural extension model in the United States?

1. The agricultural extension model has changed considerably since its origin in 1911, in response to alterations in its environment. These adjustments are one reason for its survival and relative success.

2. The agricultural extension model is based on client participation in identifying local needs, program planning, and evaluation and feedback.

3. Agricultural research activities are oriented toward potential use of research results, by use of a reward system for researchers. This utilization policy facilitates the linking function of the extension specialist and the county agent.

4. State-level extension specialists are in close social and spatial contact with agricultural researchers and professors in their specialty, and this facilitates their performance in linking research-based knowledge to farmer problems.

5. The agricultural extension model may have been more effective in its diffusion of agricultural production technology to farmers than in its latter-day diffusion of other subject matter to nonfarm audiences.

6. The agricultural extension model includes not only a systematic procedure for the diffusion of innovations from researchers to farmers, but also institutionalized means for orienting research activities toward users' needs; thus the land-grant college/agricultural experiment station/extension service complex is a total research utilization system, including innovation-diffusion as only one component.

7. If success if measured only by continued growth in funds and personnel, the U.S. extension services have been highly successful because of (1) their ability to adjust to environment changes, and (2) the strong support of the American Farm Bureau Federation and elite farm leaders.

8. The extension service's emphasis upon production agriculture has invited criticism for a lack of concern with rural social problems, some of which resulted from the activities of the extension services.

If the first county agent in 1911 could meet his counterparts today in Broome County, New York, we wonder if he would recognize their activities as representing extension work. Low-income aides? Long-distance phone calls to state extension specialists in Ithaca? Golden hamsters instead of baby beeves? The interest in home canning tips? We doubt it.

Transfer of the Agricultural Extension Model to the Third World

The agricultural extension model was particularly effective in bringing about an agricultural revolution in the United States in the period just after the end of World War II. A number of important agricultural innovations became available: antibiotic swine feeds; diethyl stibesterol for cattle; 2, 4-D weed spray; chemical fertilizers; new machines; and improved crop varieties. The agricultural extension services effectively conveyed these innovations to farmers, thus contributing to increased farm productivity. The number of consumers supported by one U.S. farmer began to climb from twelve to fifteen to twenty to thirty.

During the 1950s, government officials sought to aid the rate of economic development in the Third World nations of Latin America, Africa, and Asia. Quite naturally, it was thought that agricultural development overseas could be facilitated with the same agricultural extension model. Accordingly, U.S. extension experts served as development advisers in Latin America, Africa, and Asia during the 1950s and 1960s. They established foreign facsimiles of Iowa State, Ohio State, Cornell, and Wisconsin in Third World nations. These overseas agricultural universities were devoted to teaching agriculture and to conducting agricultural research. But in most cases the third leg of the college/research/extension program was lacking. Extension services in most Third World nations were either (1) missing entirely, or (2) organized under a national ministry of agriculture, without an effective connection to the agricultural universities. The net result was an agricultural extension system that had a certain degree of farmer contact, but not much to say to the farmer, nor much influence on the researcher.

Actually, an even more serious problem was lack of adequate farmer contact. Extension budgets were simply not adequate in the Third World, with the result that most local extension agents were responsible for 10,000 or more farmers. An effective interface across such an impossible client ratio was simply impossible. Even worse was the socioeconomic dissimilarity between the extensionists and their clients; the former were usually university graduates in agriculture, and the latter were mainly illiterates. This gap was just too wide to be bridged by attempts at effective communication (Rogers 1983). As a result, local extension agents in Third World nations tended to contact the socioeconomic elites among their clients, farmers who were most like the extension agents. This tendency contributed toward greater socioeconomic inequality in the distribution of benefits of agricultural development programs.

One solution to the problem of inadequate farmer contact was proposed by Daniel Benor, an Israeli consultant to the World Bank: the Training and Visit system (also called the T&V system)(Benor and Harrison 1977). This approach usually called for intensive communication of local extension agents with "contact farmers" (opinion leaders). The T&V system emphasized extension/farmer contact by stressing a planned, disciplined approach for extension services in Third World nations. The tremendous resources of the World Bank were placed behind the T&V system, and today over seventy-five T&V programs are under way in the Third World.

Unfortunately, the World Bank approach is widely criticized by many Third World officials, who disparage it as the "talk and vanish" system.

Only a very few evaluations of the effectiveness of the T&V system have been conducted. One of these, by Mathur and Pandey (1983) in Rajasthan, India, concluded: "The farmers had unfavorable attitudes toward the system, whereas the Village Extension Workers had a neutral attitude. . . . The criteria as envisaged in the system were not properly adopted for selecting contact farmers."

We conclude that the agricultural extension model, which achieved a certain degree of success in the United States, was only partially transferred to Third World nations. Certain of the eight key elements that we identified in a previous section were implemented in Latin America, Africa, and Asia, and others (like the research/extension interface) were not. The net result was a relatively ineffective diffusion of agricultural technology. Part of the reason for this problem lay in the particular socioeconomic conditions in the Third World. The agricultural extension model was ineptly transferred because it was not adequately understood by those who transferred it.

The county extension agent in the United States was a product of commercial agriculture, not subsistence farming. Until U.S. agriculture began to modernize, there was not so much need for an extension service. Subsistence (precommercial) farming in developing countries has not embraced the agricultural extension model with much success, a fact that suggests that the successful introduction of a knowledge transfer system must be carefully timed so that a feeling of need for its services exists or can soon be developed.

Conclusions

In this author's opinion, the agricultural extension model in the United States was transformed relatively effectively over the years since 1911, considering the vastly changed social, economic, and political conditions of the United States. But this model was transferred to the Third World nations of Latin America, Africa, and Asia without adequate modification. These poor nations could not afford a sufficient number of extension workers per thousand farmers. Nor did they effectively connect agricultural research with extension. These deficiencies, along with numerous other factors (Rogers et al., 1984), contributed toward the general lack of success of agricultural development in the Third World.

The present analysis suggests that social and agricultural transformations have been necessary ingredients in technology transfer systems—across time and across cultures. Such transformations should be considered fundamental in determining the relative success of the agricultural extension model in its various applications.

Note

I acknowledge the help of Dr. J. D. Eveland of the National Science Foundation and Dr. Alden O. Bean of Lehigh University, collaborators in identifying lessons learned from the agricultural extension model (Rogers et al. 1984).

References

Benor, Daniel and James Q. Harrison. *Agricultural Extension: The Training and Visit System.* Washington, D.C.: World Bank, 1977.

Hightower, Jim. *Hard Tomatoes, Hard Times; a Report of the Agribusiness Accountability Project on the Failure of America's Land Grant College Complex.* Cambridge, Mass.: Schenkman Publishing, 1973.

Mathur, P. N. and S. N. Pandey. "Communication in Training and Visit Extension System." *Interaction* 1 (1983): 68-76.

Rice, E. B., *Extension in the Andes: An Evaluation of Official U.S. Assistance to Agricultural Extension Services in Central and South America.* Cambridge, Mass.: MIT Press, 1974.

Rogers, Everett M. *Diffusion of Innovations*, 3rd ed. New York: Free Press, 1983.

Rogers, Everett M., J.D. Eveland, and Alden S. Bean. *Extending the Agricultural Extension Model.* Lanham, Md.: University Press of America, 1984.

von Blanckenburg, Peter. "The Training and Visit System in Agricultural Extension: A Review of First Experiences." *Quarterly Journal of International Agriculture* 21 (1982): 6-25.

CHAPTER 7

Linking Scientific and Indigenous Agricultural Systems

D. M. WARREN

Interactions between agricultural scientists, agricultural extension staff, and indigenous[1] small-scale agricultural producers have increased in recent years. Such interactions have been the result of a variety of forces, and they have been facilitated in a number of cases by formally recording the indigenous agricultural knowledge upon which cropping or animal husbandry behavior is based. These formalizations have led to an improved understanding and appreciation by agricultural scientists and extension staff of the agricultural systems developed over generations by various ethnic groups. This chapter surveys several of the reasons for an early lack of interest by scientific agriculturalists in indigenous agricultural knowledge and practices, why such an interest has begun to develop in the past few decades, and the potential for this growing interest to help research and extension improve national agriculture and the quality of life for small-scale agriculturalists. The term "small-scale producers" in this chapter refers to a variety of categories of farmers including those labeled as small-holder agriculturalists and limited-resource farmers, both subsistence farmers outside of the formal market system and emergent farmers who market part of their farm outputs.

The Evolving Nature of Interactions between Agricultural Scientists and Small-Scale Producers (SSPs)

During the past two decades there has been a growing concern for the SSP by agricultural research and extension. This interest is a dramatic departure from the past, which saw little attention given to these farmers, especially in the early days of agricultural research and extension work in many areas of the developing world.

Although it is difficult to generalize, there are a number of factors that help explain both the earlier lack of attention to the SSP and the more recent growing interest in this same population. There are also some striking similarities between the United States and the developing countries in this evolution of changing patterns of interaction between agricultural research and extension with the SSP. Some of the specific examples I will use to demonstrate the situation in the developing nations are taken from my experience in the early 1980s in Zambia where I served as team leader for the United States Agency for International Development-sponsored Zambia Agricultural Training, Planning and Institutional Development Project.

A key factor militating against interest in the SSP in the early days can be traced to negative Euro-American attitudes about non-Western peoples. Some of this attitude was generated by nineteenth-century social science literature. I have surveyed and documented these writings and attitudes and have shown how their influence is still evident in both current publications and Western folk beliefs about non-Westerners (see Warren 1976; 1979a). Many influential nineteenth-century British and American sociologists and anthropologists delineated humankind along a unilineal evolutionary ladder based on social Darwinism. In some models, evolutionary stages were defined according to racial criteria. Black populations were identified as the lowest stage of human development, labeled the savage or primitive stage. An intermediate stage of development—sometimes labeled barbaric—was represented by the so-called brown, red, and yellow peoples of the world. The highest stage of development—termed civilized—was represented by the white populations. Savages were posited to have primitive minds with intellectual endowments equivalent to European children. They were also characterized as speaking primitive languages with limited vocabularies and few abstract terms. It was believed that such populations could be assisted in their evolutionary movement toward civilized status through a Western-oriented formal education. Such ideas helped justify colonial controls over vast parts of the world and the extraction of Third World resources.

Despite the numerous studies by social scientists during the twentieth century which have shown the complex nature of patterns of social organization and of many ethnic knowledge systems and language systems, nineteenth century Western folk beliefs about non-Western populations continue to influence both the general public and many Western scientists as well. The continuing use of the concept of "traditional" versus "modern" sustains in many Western minds the notion of a dichotomy—with "tradition" connoting a simplistic, static system devoid of curiosity or intellectual exchange. Linguistic research, in fact, has revealed numerous vocabulary items for change processes and change concepts in a variety of West African languages (Warren 1979a). Yet the common attitude that SSPs have little to offer the West in terms of agricultural knowledge and much to learn from scientific agriculture has been difficult to influence.

In addition to the negative attitudes toward SSPs, another reason for an earlier lack of attention to small-scale farmers, both in the United States and the Third World, is that agricultural experiment and research stations have had a tendency to respond to the interests and requests of the better-organized farmer groups who could influence government expenditures for such research. In the United States and Canada, although there have been a number of programs designed to respond to the needs of limited-resource farmers, most of the agricultural research and extension activities have been focused where there is potential for a higher economic return, such as in the larger family farm units or in a variety of agribusiness enterprises. More efforts to delineate the particular constraints and needs of various black, Hispanic, and native American farmers in order to gain a better understanding of their indigenous agricultural knowledge and cropping systems would facilitate many of the limited-resource farmer programs sponsored through Agricultural Canada, through the United States Department of Agriculture (USDA) and through the Cooperative Extension Services of the individual U.S. states.

In developing countries there has been a strong relationship between research stations and the commercial farm sector. In those regions without a large expatriate commercial farm sector, West Africa for example, the first research stations focused their efforts on cash crops such as cocoa, rubber, and oil palms. These crops were grown on state plantations and by small-holders, and marketed through colonial and expatriate commercial enterprises.

In many former colonies, agricultural policies were set to the distinct disadvantage of indigenous SSPs. Until it gained independence in 1964, Zambia (formerly Northern Rhodesia) reserved by statute the best agricultural land for expatriate commercial farmers—particularly land along the railway line. The expatriate farmers were also paid higher prices than the African farmer for crops such as maize. Other disadvantages to the SSPs and distortions in the agricultural system were generated in Zambia's colonial era. Because of a need for cheap labor in the copper mines, authorities imposed a "native tax," payable only in cash, which forced large numbers of Zambian men to work in the mines for wages. The result to this day in a number of southern African countries, including Zambia, is a large number of farm households headed by women—as high as 40 percent in some countries. Tax revenues from the mines were invested in the urban areas occupied by the European populations, with rural areas, African agriculture, and education neglected. At independence, Zambia's copper prices were still high. There was little concern within the government or among the general population about the consequences of depleting the new nation's mineral resources; Zambia was self-sufficient in food. The new national development plans had a distinct bias toward the development of industries rather than agriculture. There was no perception of a serious problem looming in agriculture in the near future. Agriculture would take care of itself.

The impact of the colonial investment pattern on domestic education is highlighted by the fact that Zambia had only ninety-seven university graduates in all fields at independence. Although the University of Zambia opened shortly after independence, there has been a lag in the development of curricula for its school of agriculture. The B.S. in agricultural economics was not inaugurated until the 1987-1988 academic year, some twenty years later. It will be another decade before the majority of the teaching staff in the school of agriculture are Zambians. Although the government has made great strides to encourage the study of agriculture in the Zambian schools, the school system still maintains a white-collar bias, and students still seek to distance themselves from the "uneducated rural population."

In the agricultural research field, virtually all research staff were expatriates at independence. Despite a large investment by the government and various donors to train Zambian researcher and strengthen the agricultural research establishment, the majority of the professional staff remains expatriate. This fact creates two obvious constraints to interaction between Zambian agricultural researchers and SSPs. First, most Zambian agricultural lecturer/researchers at the university, as well as the Zambian senior staff at the agricultural research stations, were trained abroad and have returned with agricultural models, approaches, and attitudes conducive to interacting with commercial farmers rather than SSPs. Second, the large component of expatriate staff—characterized by frequent turnover—has more homophyly with the commercial farm sector than with the SSPs.

One distinct difference between the United States and developing countries such as Zambia is the fact that U.S. research and extension staff usually come from rural backgrounds and share some of their clientele's perspectives. Although the trend in Zambia is beginning to modify with universal primary education, staff in general tend to come from the large urban areas, and lack practical farming experience. They relate more easily to the urban elite and the commercial farm sector than to the SSP.

A distinct similarity between the United States and the developing countries has been the tendency—admittedly changing slowly in both places—for males to focus on agriculture and females on home economics in the land-grant colleges in the United States and in the agricultural colleges in the developing countries. These divisions by gender were established during colonial days and have continued into Zambia's independence even though an average of 25 percent of the small-scale farm units are female-headed (ranging from 10 to 40 percent across the country). In 1982, only 98 of the 1,387 extension officers were female (Norrby 1983b), and many of those women who were trained as agricultural extension officers were actually assigned duties as home economics extension officers—a job for which they had no formal training (Norrby 1983a).

Despite the forces preventing significant interaction between agricultural researchers and SSPs for many years, a combination of factors began to change this situation within the last two decades. These factors will be discussed in the next section.

The Changing Role of Social and Technical Scientists

Complex forces emerging on the world scene have resulted in more attention being given to the SSP by the research and extension establishments in both the West and the developing countries. An imprecise but convenient year on which to base this change is 1960, when dozens of former colonies around the world moved toward independence. With independence came training programs for local people who were expected to replace expatriates in the job market. The desire to emphasize the importance and value of indigenous cultures and languages was further stimulated by the establishment of national ethnic studies centers.

With the independence movement in full swing, there was rapid growth and expansion of the international donor community. Multilateral and bilateral aid organizations began operating in many sectors, including agriculture and rural development. National volunteer organizations such as the Peace Corps, Canadian University Service Overseas (CUSO), and the British Voluntary Service Overseas (VSO) began sending thousands of young people from the West into rural areas around the world. Many acquired and maintained a strong interest in the developing world. Hundreds of such volunteers married locally, developing a new and complex set of personal and familial relationships between the West and the developing countries. Many former volunteers have now joined the staffs of donor agencies and infused them with new attitudes about rural populations.

The new network of international agricultural research centers began in the 1960s. Social scientists began to join technical scientists on interdisciplinary teams that focused on improving agricultural productivity in developing countries. In some instances they worked actively on SSP problems.

The expectations of the populations in the newly independent countries were high. New political pressures now came from rural constituencies represented for the first time in national assemblies, requiring governments to reallocate national resources more equitably between the urban and rural sectors. The high population growth rate in many developing countries (currently about 3.2 percent per year in Zambia) resulted in national populations doubling every twenty to twenty-five years. This put many of the new governments in the serious bind of attempting to satisfy the demand for improved physical and social infrastructures at the same time that foreign exchange was becoming scarcer in many countries. The continuing trend of rural to urban migration also resulted in a decline in the number of farmers expected to supply food to an ever-growing urban, as well as general, population. In some countries scarce foreign exchange was used for the first time to import food. Agricultural policies were designed to encourage increased food production to achieve both self-sufficiency and to provide export crops to earn the foreign exchange that could be used to finance the development programs of the country.

With local nonrenewable resources such as minerals becoming depleted, some countries began to realize that labor—and, in some instances, land—was still abundant; they also began to view the SSP as an underutilized national resource. For example, in Zambia, in 1980, a majority of the approximately 730 commercial farms (those exceeding 40 hectares in size) were operated by expatriates and produced 40 percent of the marketed output on 5 percent of the total cultivated land. Another 144,000 farms (ranging in size from one to forty hectares) were operated by African households that had moved into the cash economy since independence, producing the balance of the marketed output on 15 percent of the total cultivated land. Another 462,000 farm units operating on 75 to 80 percent of the cultivated land remained subsistence farmers outside of the formal market system. Zambia's food imports have continued to rise dramatically in recent years, and its national agricultural policy now emphasizes the need to work closely with the SSP. If each subsistence farm unit could be encouraged to grow several bags of marketed produce each year, Zambia could reduce its food dependency (Republic of Zambia 1981, 16).

A dramatic development regarding SSPs has occurred in Zambia since independence. Because of food and foreign exchange shortages, many crops (sorghum, millet, cassava) grown primarily by SSPs have become import substitutes for national staples such as maize. Such cash crops as confectionary groundnuts, sunflowers, cotton, and Burley tobacco have also been grown by these SSPs for both internal use and export. Clearly, SSPs have responded to incentives for increased production. Furthermore, the vast number of subsistence farmers in the rural areas now represents a tremendous resource to tap.

Important attitude shifts toward SSPs also took place when independent countries and their educational systems challenged the Western folk beliefs about non-Western peoples. Volunteers from the West, some already fluent in languages of the developing countries, returned to area studies programs in their home universities to further their knowledge of their former host countries. Many have now conducted formal field research in developing countries, greatly adding to the knowledge base on those countries.

In the field of anthropology, an important development began in the early 1960s known variously as cognitive anthropology and ethnoscience. The new set of field methodologies allowed the anthropologist to gain a better understanding of how a culture perceives its universe through its language. The approach requires an in-depth knowledge of the local language to record formally just how people use language to define their physical, social, and intellectual environments and how these knowledge systems enable the decisionmaking process to function. This kind of analysis might explain how a Chewa farmer in Zambia defines and classifies soil types and crop varieties, and how these indigenous classification systems lead to particular crops being planted on particular types of soils. The fact that many of these indigenous classification systems differ greatly from the scientific systems has important implications for the exchange of information between SSPs and agricultural research and extension staff. As numerous studies of this type have become available, it is apparent that many indigenous knowledge systems are very complex, but rational and functional. By formally recording such systems, we can improve understanding and communications between change agents and local populations.

Other changes in attitude toward SSPs came through various donor agencies. USAID, established in 1962, made a variety of program and philosophical changes in the early 1970s. In 1973

> the United States became the first of the rich countries to shift away from . . . "trickle-down" [economics and information dissemination approaches] to people-oriented development. Redirecting our efforts toward the poor who make up the overwhelming majority of people in the developing countries conforms to the instinctive concern of Americans that past foreign aid programs have not reached the people we have wanted to help. (Owens and Shaw 1974, xiii)

The 1973 legislation became known as the New Directions to U.S. Development Assistance, giving the highest priorities to projects that improve the lives of the poorest segment of society and increase their ability to participate in national development (Mickelwait et al. 1979, 2-3). Also in 1973, the Percy amendment called for development efforts that would improve the status of women through USAID projects. In 1974, USAID began to require social-soundness analyses for all project design and evaluation documents. These analyses were to measure the compatibility of development projects with the sociocultural environment and to estimate both the spread effect of projects and the social impact of the project in terms of the distribution of benefits and burdens (USAID 1978). Social science positions were established in USAID/Washington and its regional offices to assist in such analyses. The requirement for host country participation in project design, implementation, and evaluation coincided in several developing countries with new efforts to decentralize planning and budgetary functions from the central government out to the district level. Decentralization programs have made government officials more accountable and more responsive to local populations, including SSPs.

A final change occurring in the 1970s was the Farming Systems Research and Development efforts pioneered by a variety of individuals, such as David Norman. These programs involve interdisciplinary teams of social and technical scientists and the active participation of the farmer in on-farm trials. The goal is to identify the constraints faced by the SSPs and focus research efforts on those problems.

An Assessment of the Impact of Research on Indigenous Agricultural Knowledge Systems for Improving Agricultural Development

Despite all of these forces, the reorientation of agricultural research and development programs toward a closer interaction with SSPs has emerged slowly. Nevertheless, a survey of some of the more recent efforts to improve the linkage between scientific and indigenous agricultural systems indicates steady movement in that direction. One accomplishment has been the design of the perceptual grid, a pedagogical tool that outlines the different sets of blinders an individual acquires—usually unconsciously—through his or her enculturation and formal education and which cause people from different backgrounds to perceive phenomena differently, thus making communication difficult (Warren and Patten 1981). It also explains what happens to an individual who has crossed a language and cultural barrier and gone through the acculturative process. These are important lessons to be learned by anyone hoping to work effectively in an interdisciplinary team and with SSPs from a different culture.

Baviskar gives an interesting and sometimes amusing example of a Third World perspective on the interpretation of rural development efforts in the United Kingdom. He states that

> . . . it is strange how human beings learn so little from each other's experience. Many Third World countries are repeating the mistakes of industrialized countries in the process of development. There is no reason why advanced countries should not learn a thing or two from the Third World countries who have vast experience in the field of rural development. (Baviskar 1981, 8-9)

Chambers has eloquently discussed the problems inherent in facilitating cross-disciplinary communication as well as communication between scientists and SSPs. He stated that

> . . . the dominant professional values of scientists concerned with agricultural research impair their ability to work with small farmers. This statement applies to social scientists as well as to natural scientists. . . . Scientists must first understand themselves. Without self-critical introspection, scientists are likely to hold views of small farmers and their environments that are largely determined by professional training, by the ways in which scientists are exposed to small farmers, and by scientists' own environments. . . . Professional education and training, so often regarded as unequivocally beneficial, can also be seen as a lengthy process of conditioning in selective perception [so that we become aware of] our professional blinders and tunnel vision. (Chambers 1980, 1)

Chambers also comments on prejudice against indigenous knowledge:

> Modern scientific knowledge is centralized and associated with the machinery of the state; and those who are its bearers believe in its superiority. Indigenous technical knowledge, in contrast, is scattered and associated with low prestige rural life; even those who are its bearers may believe it to be inferior. It is difficult for some scientists to accept that they have anything to learn from rural people, or to recognize that there is a parallel system of knowledge to their own, which is complementary, usually valid, and in some respects superior. (Chambers 1980, 2)

Chambers goes into these matters in far greater detail in his latest book (Chambers 1984).

A recent publication by Rhoades and Booth centers on interdisciplinary experiences at the International Potato Center in Peru.

> Differences in perception and role definitions between biological and social scientists result in a mutual respect that is miserably low . . . the upshot of this disciplinary tribalism [being] that social and biological scientists tend to line up on opposite sides of the fence and throw spears. (Rhoades and Booth 1983, 2)

Attempting to improve storage losses of potatoes, however, the social scientists were able to reveal differences in perceptions between farmers and technicians over the concept of "loss" versus "waste." This improved understanding of different perceptions resulted in a positive reorientation of the experiment station's on-farm research and helped move the farmers from being passive recipients to being creative actors experimenting on their own. These "farmers with long-term practical experience with their land, mix of crops, climate, and local socioeconomic conditions assume the status of experts in their own right and members of the problem-solving team" (Rhoades and Booth 1983, 5).

Although formal studies on indigenous knowledge systems and indigenous technical knowledge have flourished only over the past two decades, there are earlier examples of individuals who were sensitive to the value of understanding such systems. An eloquent discussion of the rationale behind the Kikuyu mixed-cropping system was made by Leakey (1936). He provided a lengthy explication of the many positive aspects of such a cropping system—all of which were overlooked by the Europeans. He compared the mixed cropping with the negative impact of European monocropping in the Kikuyu environment. He argued that

> The habit of regarding African methods of agriculture or of any other activities as inherently bad because they are different from our own is most unwise. I do not suggest that the methods used by different native tribes are all perfect. Doubtless the methods of agriculture employed by the Kikuyu could almost certainly be improved in many details, but this could only be done if European methods of research were employed in trying to develop the African method of cultivation, which is a very different thing from trying to substitute European methods of planting for those which have been evolved out of research by trial and error. (Leakey 1936, 122)

For technical agricultural scientists unfamiliar with the literature on indigenous agricultural systems, there is a variety of valuable sources. Polly Hill, an agricultural economist, has made an excellent "plea for indigenous agronomics" (Hill 1970). Brokensha, Warren, and Werner (1980) have edited a text on the subject, with contributions from Glenn Custred on the role of indigenous knowledge systems in the design of national agricultural census questionnaires, from Steve Brush on the role of indigenous potato taxonomies in allowing scientists to understand potato-cropping behaviors, and from Christina Gladwin on discovering the indigenous criteria used by SSPs in deciding not to adopt a Plan Puebla agronomic recommendation on fertilizer application. Another excellent source is Peggy Barlett's (1980) edited text *Agricultural Decision Making.*

> [Kurin's] symbolic analysis of Punjabi agriculture in Pakistan's Indus Basin reveals an anthropomorphic humoral agronomy used by farmers to make crop production decisions. This agronomy is found to be systemic, coherent, and empirically constituted. . . . The cultural analysis of the system of agronomical knowledge used by Pakistani peasant farmers reveals the rationale for the ways in which inputs of the green revolution have been accepted, modified, or rejected, [an explication of] the indigenous conceptual scheme for agronomical decision making. (Kurin 1983, 283)

Although there is a tendency to let such studies stand on their own—usually on bookshelves—it is possible to introduce them and their underlying concepts into training programs where their impact can be considerable. I was able to assist in the development of training materials for the Economic and Rural Development Management (ERDM) project sponsored by USAID in Ghana. Working with regional Ghanaian teams of trainer-consultants, we developed an annual management/development planning in-service training program for government officials in all sixty-eight districts and nine regions. We developed group exercises to sensitize officials to cross-disciplinary problem identification and problem-solving approaches and to improved inter- and intraethnic communications. The exercises were also developed to help officials appreciate the value of indigenous knowledge systems and to identify local resources for development, including indigenous decisionmaking organizations of farmers and market sellers. We also worked to discover and understand local perceptions of development issues, such as taxation. These exercises are described in detail in Warren (1986), and many of the actual exercises are available in Warren (1982) and have been used in Iowa State University's Development Advisory Team (DAT) Training Program. One of the goals of the program is to train Iowa State faculty and staff working as multidisciplinary development advisory teams to conduct international development project design and evaluation assignments in an effective and efficient manner. Another is to train them to work effectively and efficiently in a multicultural, cross-disciplinary team by using techniques that foster cooperation, and coordination. More than 520 participants have attended twelve training sessions in the past six years, indicating a strong interest and desire by many professionals to improve their capacity to work in both cross-disciplinary and cross-cultural settings.

By working with and through indigenous decisionmaking groups, important cost-effective development results can be achieved. Based on the ERDM training program in Ghana, two important local development efforts emerged. The first involved indigenous healers who were identified as a resource for improving health conditions in areas where the Ministry of Health lacked resources to extend its services. The Primary Health Training for Indigenous Healers (PRHETIH) program was begun in Techiman District. The results have been very positive, and the program has now been extended into other districts (Warren 1979b; Warren et al. 1982). The second development effort grew out of a better understanding of local attitudes toward taxation in Techiman District. New and more appropriate district revenue collection methods were devised locally, resulting in a 311 percent improvement in tax collection in three fiscal years. The indigenous market unions were recognized as important organizations and brought into the decisionmaking process for district development; this new involvement increased collected market fees by 422 percent in three years. As a result, the amount of revenue allocated for capital improvements in the market was increased from 20,000 in 1979/80 to 300,000 Ghanaian cedis in 1981/82 (Warren and Issachar 1983). Other studies have demonstrated the complexity of indigenous aesthetic knowledge systems and their role in the development of the Ghanaian crafts industry (Warren and Andrews 1977).

In an era of scarce resources, it is more cost-effective to build development programs for agriculture and rural development on indigenous decisionmaking organizations than to create new organizational structures that would have to be imposed from outside. It is also cost-effective to take the time and effort to delineate the indigenous agricultural technical knowledge and the indigenous agricultural decisionmaking systems upon which agricultural production practices are based. By doing so, one is better able to understand where isomorphism exists between the local and scientific system, where communication difficulties may occur between the two systems, which aspects of the indigenous system may be superior to proposed innovations, and which could be improved by working with and through the existing system. One also has a far greater insight into how the local farmer assesses risk. One understands better, for example, the reluctance of many SSPs in Zambia to adopt the SR52 hybrid maize variety being encouraged through a national extension program. The SSP recognizes both production and marketing risks in adopting this hybrid variety, particularly given the occasionally unreliable governmental agricultural support system in the rural areas. The market for quality seed supply is sometimes unstable. (For example, one year in the early 1980s, SR52 seed with a 26 percent germination rate was delivered in one district.) The variety is highly subject to weather variations and is particularly susceptible to drought. It requires chemical fertilizer, which frequently must be obtained on credit. It also presupposes an effective extension system that is going to ensure that the correct amount of fertilizer is spread in the correct place at the correct time. The SR52 variety of maize is also susceptible to weevils and cannot be stored well in the indigenous grain storage structures. Also, it must be marketed through official channels. Because maize breeders now recognize these local risks, a new maize-breeding program has been undertaken to improve the local open-pollinated varieties. There are also efforts under way to revitalize the indigenous Chewa concensus decisionmaking group (the *mphala*) to facilitate SSP development (Douglas 1983).

It is clear that we can learn from the SSP about local plant and animal disease and pest management, about ecological relationships between different types of soils and crops, about the advantages of mixed and multiple cropping, and about the decisionmaking modes of local management groups.

Key Current Trends and Issues Linking Scientific and Indigenous Agricultural Systems

One encouraging trend over the past decade has been the increase in the number of agricultural researchers and extension experts who recognize the value of understanding the SSP. A. Colin McClung in his foreword to Harwood (1979) notes that "researchers must be unusually adept at seeing the world from the farmer's vantage point" (xi). Harwood himself believes that "the purposeful blending of traditional and modern technologies may well prove the key to starting the most disadvantaged farmers along a more rapid development path" (xiii).

> Increasing interest in these farm-centered research systems has led scientists in the cropping systems program at the International Rice Research Institute to make a thorough appraisal of the knowledge and resources to be found in traditional farming systems. In 1972, they began to document some of this centuries-old knowledge and to measure the efficiency of resources used in these traditional systems. The result of these investigations has been a greatly increased respect for the traditional farmer and a new effort to combine traditional farming knowledge and skills with the trained insights and experimental method of the scientific researcher. (Harwood 1979, 34, 36)

Another example is found in Shaner, Philipp, and Schmehl (1982), who describe how the farming systems research (FSR) team is expected to understand "the farmers' knowledge of alternative management practices, cropping patterns, sources of inputs, information, and markets" (65). And Compton (1983) provides excellent insights into the role of indigenous knowledge systems for improving the capacity of the extension worker to communicate with the SSP.

Despite improving attitudes toward the SSP by scientific researchers, there are still important steps that must be taken. To date, we have a large number of empirical studies explicating indigenous knowledge systems and a number of projects that are working through these systems. Most of these studies, however, do not explain how to go about actually recording such systems formally. There are a number of methodological approaches, some of which are outlined in Chambers (1984) and in MARC (n.d.), but a manual is needed. Compton, Paul Richards, and I have begun to develop such a manual for use in agricultural colleges and extension training programs for degree and diploma students in agricultural research and extension. It would also be useful for members of farming systems research teams whose composition may be interdisciplinary but who lack direct experience in recording such systems. In Zambia such a manual could also be used by the new adaptive research and planning teams that are now based in each provincial capital and conducting their research along FSR approaches. These teams, composed mainly of agricultural economists, extension specialists, and agronomists, would benefit from a clear set of guidelines to assist them in tapping into the various indigenous agronomic knowledge systems that exist across Zambia.

Two programs that were being actively considered with officials of the University of Ghana and the Ghanaian extension training system have been postponed until a more stable politico-economic environment is achieved. One program was to incorporate material into the extension training syllabi about the importance of indigenous knowledge systems as national resources for extension workers operating in the field. Agricultural extension trainees would have been expected to conduct practica in which they would record indigenous agronomic systems. Over a period of several years it would be possible to generate formally recorded soil classification systems for all of the sixty ethnic groups within Ghana. If the soil types were recorded, along with the criteria used by farmers to differentiate and label these soil types, and these indigenous soil classifications were compared with the scientific soil classification system used by extension staff, potential communication problem areas between extension staff and SSPs could be anticipated. Extension workers from one ethnic group posted to work with farmers from a different ethnic group would have available a ready-reference document. Other knowledge domains to be studied could include soil-crop relationships, crop varieties, pest management, and concepts of volume, space, and time.

A second program would have made the results of these practica the basis for a national indigenous knowledge resource center where the data could be maintained and made available to any interested person. It would be a systematic approach to recognize, record, and preserve these important and usually overlooked national resources. (Similar exercises could be conducted in the United States where there is still a need to record the agronomic knowledge systems of different ethnic groups.) If such documentation programs could be initiated in various developing countries, the important task of making communications easier between SSPs, extension officers, and agricultural research staff would be greatly facilitated. Such programs would also help to sensitize administrators and agricultural planners and policymakers to better realize the importance of working with SSPs and establishing a two-way communication system between them and the professional agricultural staff. Such efforts should also enhance the many important contributions made by the national and international agricultural research centers and by the numerous farming systems research teams.

Notes

The author would like to thank Edward Rawson, Randy Benoit, Patrick Ncube, Constantina Safilios-Rothschild, and John Douglas for useful comments and criticisms of drafts of this chapter.

1. The term "indigenous" is used to refer to a localized agricultural system that has developed over time with cropping patterns based on an agricultural knowledge system expressed in the local language; such a system is viewed to be in a dynamic equilibrium with the environment, influenced by innovations emerging from within the system as well as those adopted from other indigenous systems and from the national and international agricultural systems. The term "traditional" has not been used in order to avoid the Western popular connotation that the system is static and simple even though recent research has shown the dynamic, complex nature of such systems.

References

Barlett, Peggy F., ed. *Agricultural Decision Making*. New York: Academic Press, 1980.

Baviskar, B.S. "Rural Development in the United Kingdom: A Third World Perspective." IDS discussion paper. Brighton, England: Institute for Development Studies, University of Sussex, 1981.

Brokensha, David, D. M. Warren, and Oswald Werner, eds. *Indigenous Knowledge Systems and Development*. Lanham, Md.: University Press of America, 1980.

Chambers, Robert. "Understanding Professionals: Small Farmers and Scientists." IADS occasional paper. New York: International Agricultural Development Service, 1980.

——. *Rural Development: Putting the Last First*. Harlow, England: Longmans, 1984.

Compton, J. Lin. "Linking Scientist and Farmer: Rethinking Extension's Role." In *Farming Systems in the Field*. Proceedings of Kansas State University's 1982 FSR Symposium. Farming Systems Research Series, Paper No. 5, Cornelia Flora, ed. Manhattan: Kansas State University, 1983, 229-40.

Douglas, John. *A Plan for Helping Village Churches Reach Economic Self-Sufficiency Through Agriculture*. Chipata, Zambia: Church of Christ, 1983.

Harwood, Richard R. *Small Farm Development: Understanding and Improving Farming Systems in the Humid Tropics*. Boulder, Colo.: Westview Press, 1979.

Hill, Polly. "A Plea for Indigenous Agronomics: The West African Example." In *Studies in Rural Capitalism in West Africa*, Polly Hill, ed. Cambridge: Cambridge University Press, 1970, 3-20.

Kurin, Richard. "Indigenous Agronomics and Agricultural Development in the Indus Basin." *Human Organization* 42, no. 4(1983):283-94.

Leakey, L. S. B. "Science and the African." In *Kenya: Contrasts and Problems.* London: Methuen, 1936.

MARC. *Some Methodological Issues in the Measurement, Analysis and Evaluation of Peasant Farmers' Knowledge of Their Environment.* MARC Report No. 9. London: University of London, Chelsea College, Monitoring and Assessment Research Centre (MARC), n.d.

McClung, A. Colin. *Foreword to Small Farm Development*, by Richard R. Hardwood. Boulder, Colo.: Westview Press, 1979.

Mickelwait, Donald R., Charles F. Sweet, and Elliott R. Morss. *New Directions in Development: A Study of USAID.* Boulder, Colo: Westview Press, 1979.

Norrby, Gunnar. "Action Plan" for Manpower Development of Field Staff for the Department of Agriculture. Lusaka: Ministry of Agriculture and Water Development, 1983a.

——. *Manpower Study within the Department of Agriculture: Summary Report.* Lusaka: Ministry of Agriculture and Water Development, 1983b.

Owens, Edgar, and Robert Shaw. *Development Reconsidered: Bridging the Gap Between Government and People.* Lexington, Mass.: D. C. Heath, 1974.

Republic of Zambia. *Food Strategy Study.* Annex: Socio-Regional Framework. Lusaka: Ministry of Agriculture and Water Development, 1981.

Rhoades, Robert E., and Robert H. Booth. "Interdisciplinary Teams in Agricultural Research and Development." *Culture and Agriculture*, no. 20(1983):1-7.

Shaner, W. W., P. F. Philipp, and W. R. Schmehl. *Farming Systems Research and Development.* Boulder, Colo.: Westview Press, 1982.

USAID. *Social Soundness Analysis.* AID Handbook 3, Appendix 4A. Washington, D.C.: USAID, 1978, 1-12.

Warren, D.M. "Indigenous Knowledge Systems for Activating Local Decision-Making Groups in Rural Development." In *Communications for Group Transformation in Development*, Godwin Chu, Syed Rahim, and D.L. Kincaid, eds. Communications Monograph No. 2. Honolulu: The East-West Center, 1976, 307-29.

——. "Humanistic Approaches to Applied Anthropology." In *Essays in Humanistic Anthropology*, Bruce Grindal and D.M. Warren, eds. Lanham, Md.: University Press of America, 1979a, 115-35.

——. "The Role of Emic Analyses in Medical Anthropology." In *African Therapeutic Systems*, Z. A. Ademuwagun, J. A. A. Ayoade, I. E. Harrison, and D. M. Warren, eds. Los Angeles: Crossroads Press, 1979b.

——. *Development Advisory Team (DAT) Training Program Manual.* Ames: Iowa State University/USAID Title XII Strengthening Grant Program, 1982.

——. "Anthropology and Rural Development in Ghana." In *Anthropology and Rural Development in West Africa*, Michael M. Horowitz and T. Painter, eds. IDA Monographs in Development Anthropology, No. 1. Boulder, Colo.: Westview Press, 1986, 63-91.

Warren, D.M., and J. Kweku Andrews. *An Ethnoscientific Approach to Akan Arts and Aesthetics.* Working Papers in the Traditional Arts, No. 3. Philadelphia: Institute for the Study of Human Issues, 1977.

Warren, D.M., S. Bova, M. A. Tregoning, and M. Kliewer. "Ghanaian National Policy Towards Indigenous Healers: The Case of the Primary Health Training for Indigenous Healers (PRHETIH) Program." *Social Science and Medicine* 16 (1982):1873-81.

Warren, D.M., and J. D. Issachar. "Strategies for Understanding and Changing Local Revenue Policies and Practices in Ghana's Decentralization Programme." *World Development* ll, no.9(1983):835-44.

Warren, D.M., and S. Patten. "Cross-cultural Interactions: The Transition and Reentry Processes." *Anthropology and Humanism Quarterly* 6, no. 1(1981):20-24.

PART 3

FUTURE CHALLENGES

Prospects for bringing about substantial improvement in the design and implementation of policies and programs for agricultural development in the Third World probably center on several strategic leverage points. Some of the more important ones are the development of national research capacities and the evolution and integration of farming systems research and extension philosophy and methods in overall national development programs.

Ruttan assesses the evolution of the international research system since World War II, indicating organizational and management issues that have been confronted and suggesting several perspectives to guide future policies and programs. While acknowledging the contribution of U.S. universities in training scientists from developing countries, the author questions the capability of such U.S. institutions to assist with technology-oriented research in the tropical and subtropical countries.

The final chapter by Waugh, Hildebrand, and Andrew describes the basis for the emergence of research and educational activities focused on the whole farm system, the farm family, the decisionmaking process of these small-scale producers, and their active participation in a research and development process aimed at generating appropriate technology.

CHAPTER 8

The International Agricultural Research System

VERNON W. RUTTAN

At the end of World War II, the Food and Agriculture Organization (FAO) of the United Nations was established to perform the functions of a global ministry of food and agriculture[1] The FAO headquarters were established in Rome to provide technical assistance to individual countries; support agricultural education; collect statistics on land use, food consumption, and agricultural production and trade; and publish periodicals, bulletins, and yearbooks on the world food and agriculture situation.[2]

The FAO has—through its technical assistance, educational, and regional networking activities—made significant contributions to the development of national research capacity in agriculture. But the FAO's governing council and program committee were reluctant to approve a substantial role in the sponsorship and conduct of research as part of the FAO's regular program. Lack of support for research was due in part to failure to perceive the causes of lagging agricultural development in the poor countries at the time the FAO was established. Lagging agricultural development was interpreted primarily as the failure to make effective use of available technology. The lack of more productive technology was not itself seen as a major barrier to agricultural development.

By the late 1950s, this view had changed. Technical assistance programs, based explicitly or implicitly on a technology-transfer model, failed to generate either rapid modernization of traditional farms and villages or rapid growth in agricultural production.

Between the late 1960s and the late 1970s, a new international agricultural research system was established. The development of this new system represents a dramatic example of scientific entrepreneurship. As a result of this effort, the vision of the agricultural leaders who established the FAO after World War II has been realized. The global ministry of agriculture has been complemented by a global agricultural research service.

In this chapter I attempt to place the evolution of the international agricultural research system within the context of other post-World War II agricultural development assistance programs. I also indicate some of the organizational and management issues that have confronted, or continue to confront, the management of the international agricultural research system. And I present some of my own perspectives on future policy and program directions.

The International Agricultural Research Institute Model

Experience with development assistance in agriculture over the last several decades has resulted in three models for organizing professional resources to work on problems of agricultural and rural development. These can be described as the counterpart, the university contract, and the international institute models.

The counterpart model is perhaps the most familiar. It refers to a situation in which the individual scientist, or so-called expert, is employed by a technical assistance agency to function in an advisory role in close cooperation with counterpart scientists or professionals in national research, education, or program operations. The staff member from the external technical assistance agency is typically viewed as an expert who functions in an advisory role relative to his or her national counterpart. Assistance agencies gradually came to understand that the transferability of agricultural expertise from temperate to tropical regions was severely limited. The scientific knowledge needed to improve both agricultural productivity and institutional performance could only be obtained by the development of location-specific research capacity on the part of both expatriate and indigenous scientists. As a result, by the mid-1960s, the use of foreign experts as advisers was increasingly regarded as an inadequate response to the technical and scientific assistance needs of the developing countries. Nevertheless, the counterpart approach continues to be an important way of organizing technical assistance efforts on the part of national and multilateral and technical assistance agencies. The World Bank, for example, often incorporates a substantial amount of technical assistance, organized in the counterpart model style, as part of its project lending activity.

The university contract model typically has involved the

establishment of a special relationship between a university in a developed country and a university in a developing country. At times, this model has also been employed to link a consortium of institutions in developed countries and/or developing countries. Occasionally, the link has also involved a ministry-level research division or institute. The university contract model has been employed where institution building and training have represented the major objectives of technical assistance activity. Frequently, the institution-building objective has involved, either explicitly or implicitly, positive assumptions about the relevance of the land-grant experience of the United States to the solution of problems of technical or institutional innovation in the host country.

In the past, the university contract model has rarely provided a conducive environment for the long-term commitment of professional resources to scientific and technical research on agricultural problems in developing countries. It has made an important contribution to the training of large numbers of agricultural scientists from developing countries at the postgraduate level. But the university contract model has not, with few exceptions, led to the discovery of new knowledge or the invention of the new technology needed to expand productive capacity in developing countries. A new effort has been initiated (under Title 12 of the International Development and Food Assistance Act of 1975) to establish a more adequate financial and administrative environment for the participation of colleges of agriculture in the United States in an effort to strengthen national research programs in the developing countries. Experience thus far seems to indicate that the lessons of the 1960s—that the comparative advantage of developed countries' institutions lies more in the training of scientists from developing countries than in technology-oriented research for the developing countries—will have to be relearned.

Between the mid-1960s and the early 1970s, the international research and training institute emerged, in the eyes of the international aid agencies, as the most effective way to organize scientific capacity to generate technical change for agriculture in the developing countries. The international institute model draws on two historical traditions. One is the experience of the great tropical research institutes that played an important role in increasing the production of a number of export commodities, including rubber, tea, sisal, and sugarcane. The Rubber Research Institute of Malaya was an excellent example of the colonial research institute. The sugar and pineapple research institutes in Hawaii were examples of private research institutes supported by industrial organizations.

Table 8.1. The International Agricultural Research Institutes

Center	Location	Research	Coverage	Date of Initiation	Core Budget for 1988 ($000,000)
IRRI (International Rice Research Institute)	Los Banos, Philippines	-Rice under irrigatrion, multiple cropping systems: upland rice	Worldwide, special emphasis on Asia	1960	25.7
CIMMYT (International Center for the Improvement of Maize and Wheat)	El Batan, Mexico	-Wheat(also triticale, barley): maize (also high-altitude sorghum)	Worldwide	1966	24.3
IITA (International Institute of Tropical Agriculture)	Ibadan, Nigeria	-Farming systems: cereals (rice and maize as regional relay stations for IRRI and CIMMYT): grain legume (cow-peas, soybeans, lima beans); root and tuber crops (cassava, sweet potatoes, yams)	Worldwide in lowland tropics, special emphasis	1967	21.8
CIAT (International Centre for Tropical Agriculture)	Palmira, Colombia	-Beef; cassava; field beans; swine (minor); maize and rice (regional relay stations to CIMMYI and IRRI)	Worldwide in lowland tropics, special emphasis on Latin America	1968	24.1
WARDA (West African Rice Development Association)	Monrovia, Liberia	-Regional cooperative effort in adaptive rice research among 13 nations with IITA and IRRI support	West Africa	1971	4.6
CIP (International Potato Centre)	Lima, Peru	-Potatoes (for both tropical and temperate regions)	Worldwide, including linkages with developed countries	1971	15.3

ICRISAT (International Crops Research Institute for the Semi-Arid Tropics)	Hyderbad, India	-Sorghum; pearl millet; pigeon peas; chickpeas; farming systems; groundnuts	Worldwide, special emphasis on dry semiarid tropics, non-irrigated farming.	1972	24.6
IBPGR (International Board for Plant Genetic Resources)	FAO. Rome, Italy	-Conservation of plant genetic material with special reference to crops of economic imortance	Worldwide	1974	6.0
ILRAD (International Laboratory for Research on Animal Diseases)	Nairobi, Kenya	-Trypanosoiasis; theileriasis	Mainly Africa	1973	13.1
ILCA (International Livestock Center for Africa)	Addis Ababa, Ethiopia	-Livestock production system	Major ecological regions in tropical zones of Africa	1974	17.4
ICARDA (International Centre for Agricultural Research in Dry Areas)	Lebanon, Syria	-Crop and mixed farming systems research, with focus on sheep, barley, wheat, broad beans, and lentils	West Asia and North Africa, emphasis on the semiarid winter precipitation zone	1976	18.7
IFPRI (International Food Policy Research Institute)	Washington, D.C. United States	-Food policy	Worldwide	1975	8.3
ISNAR (International Service for National Agricultural Research)	The Hague, Netherlands	-Strengthening the capacity of national agricultural research programs	Worldwide	1980	6.5

Sources: J.G. Crawford, "Development of the International Agricultural Research system,: in ***Resource Allocation and Productivity in National and International Agricultural Research***, Thomas M. Arndt, Dana G. Dalrymple, and Vernon W. Ruttan, eds. (Minneapolis: University of Minnesota Press, 1977), pp. 282-83. Budget data for 1988 were obtained from the Secretariat for the Consultative Group on International Agricultural REsearch World Bank, Washington, D.C.

The International Institute model also draws on the experience of the Rockefeller Foundation in Mexico, and the Ford Foundation and the Rockefeller Foundation in the Philippines in support of the research that led to dramatic increases in food crop production, particularly wheat and rice. The complex of international agricultural research institutes listed in Table 8.1 evolved directly from two institutions: the International Rice Research Institute (IRRI) and the International Center for Improvement of Maize and Wheat (CIMMYT), established by the Rockefeller Foundation and the Ford Foundation in 1959 and 1963.[3] The Rockefeller Foundation's agricultural sciences program was initiated in 1943 with the establishment of the Office of Special Studies (Oficina de Estudios Especiales) in cooperation with the Mexican Ministry of Agriculture. Field research programs were first initiated with wheat and corn. The program was later expanded to include field beans, potatoes, sorghum, vegetable crops, and animal production. A common pattern of staffing was followed for each commodity program. A specialist from the United States was brought in as each commodity program was initiated. Each specialist assembled a staff of young Mexican college graduates who were trained in research methods and practices. Some of the most promising participants were later selected for further training at the postgraduate level.

The wheat program was successful almost from the beginning. Stem rust was diagnosed as a major factor limiting wheat production. Resistant wheat varieties were being distributed to farmers by the fall of 1948. By 1956, the impact on productivity was sufficient to make Mexico independent of the need for imported wheat for a number of years.

The success of the wheat program in Mexico led to conversations between the Ford Foundation and the Rockefeller Foundation about the possibility of collaborating on a program in Asia. After extended investigation and negotiation, an agreement was reached between the two foundations and the Government of the Philippines to establish the IRRI in the Philippines. The agreement was signed in 1959. By 1962, the institute's facilities had been constructed, a staff was being recruited, and the program was under way.

The establishment of the IRRI as an autonomous research institute with its own governing body and its own staff departed sharply from the collaborative pattern employed by the Rockefeller Foundation in Mexico. A major reason for the success of the Rockefeller Foundation's program in Mexico was its economical use of the limited number of professionals available in Mexico at the time the program was initiated. The shortage of professionals and indigenous educational resources was conducive to the development of an internship system that ultimately linked professional education with applied research and technology development. However, in situations in which the scientific and technical problems were more complex than in the early wheat work, the patterns worked out during the early years of the Mexican program were not as effective. In more complex situations, gathering together a team of highly skilled senior scientists from several disciplines appeared to be a more appropriate staffing strategy. This was the pattern followed in the establishment of the IRRI. It was also the approach adopted in 1963 when the Rockefeller Foundation's program staff in Mexico was reorganized into the new CIMMYT to facilitate a stronger international collaborative research effort in wheat and maize.

During the late 1960s, the Ford Foundation and the Rockefeller Foundation again collaborated, establishing the International Institute of Tropical Agriculture (IITA) in Ibadan, Nigeria, and the International Center for Tropical Agriculture (CIAT) in Palmira (near Cali), Colombia. With the establishment of these two new centers, the financial requirements of the system began to exceed the capacity of the two foundations. Consultations were held between the Ford and Rockefeller Foundations, the World Bank, the FAO, and the United Nations Development Program (UNDP) in 1969. Following several informal meetings in January 1971, a formal meeting was held in May 1971 to organize the Consultative Group on International Agricultural Research (CGIAR). The initial membership included the World Bank, the FAO, and the UNDP as sponsors, plus nine national governments, two regional banks, and three foundations. Membership had grown to thirty-nine institutions by 1980.

The leadership for the CGIAR is now centered at the World Bank, which provides a chairperson and a secretariat for the group. To provide technical guidance for its work, the CGIAR established the Technical Advisory Committee (TAC). The TAC consists of a chairperson and twelve scientists. The FAO supports the TAC Secretariat. Technical matters such as new institute initiatives and program changes at existing institutes are referred to the TAC for technical review before action by the CGIAR. The TAC develops draft policy statements for the CGIAR's consideration on priorities within the system and has the authority to initiate investigations, and to suggest initiatives and program changes to the CGIAR on its own initiative. Since 1976, the TAC has been charged with the responsibility of organizing comprehensive quinquennial reviews of the programs that have common elements in the several centers, such as cropping systems or mechanization research.

The international research system has grown rapidly. Expenditures rose from only $1.1 million in 1965 to about $180 million in 1984 (Figure 8.1). The rapid growth is a result of several factors. One is the perception that research offers a high payoff and is an efficient source of agricultural growth. The intellectual basis for this perception was provided in T.W. Schultz's iconoclastic book, *Transforming Traditional Agriculture*.[4] The empirical support was provided by the rapid diffusion of both the new wheat varieties in Mexico and of the new rice varieties in the Philippines.

A second factor was the efforts of both the multilateral and the bilateral assistance agencies to find a viable alternative to the counterpart and university contract models. Although the university contract model was viewed as an effective device for building professional capacity, critics usually gave it low marks for its ability to bring that capacity to bear on the generation of more productive agricultural technology.

On this point it may be appropriate to quote my own perspective from the late 1960s:

> The United States is characterized by a highly developed institutional infrastructure linking the university to other private and public institutions involved in technical, social and economic change. In societies where such an infrastructure has developed, research and education within the framework of the traditional academic disciplines and professions have represented an effective link in a larger system devoted to the production, application and dissemination of new knowledge. . . . The American pattern of academic and professional organization, when transplanted into societies where the presumed institutional infrastructure does not exist, rarely performs as an effective instrument of technical . . . change.[5]

The productivity of public sector agricultural research is dependent on close articulation among the several public and private sector institutions that are engaged in the generation of new knowledge, in technology development, and in the dissemination of new knowledge and technology. And the development of such institutions must be consistent with the development of other political and economic institutions and with resource and cultural endowments.

Organization and Management Issues

A number of areas of concern with respect to the future of the

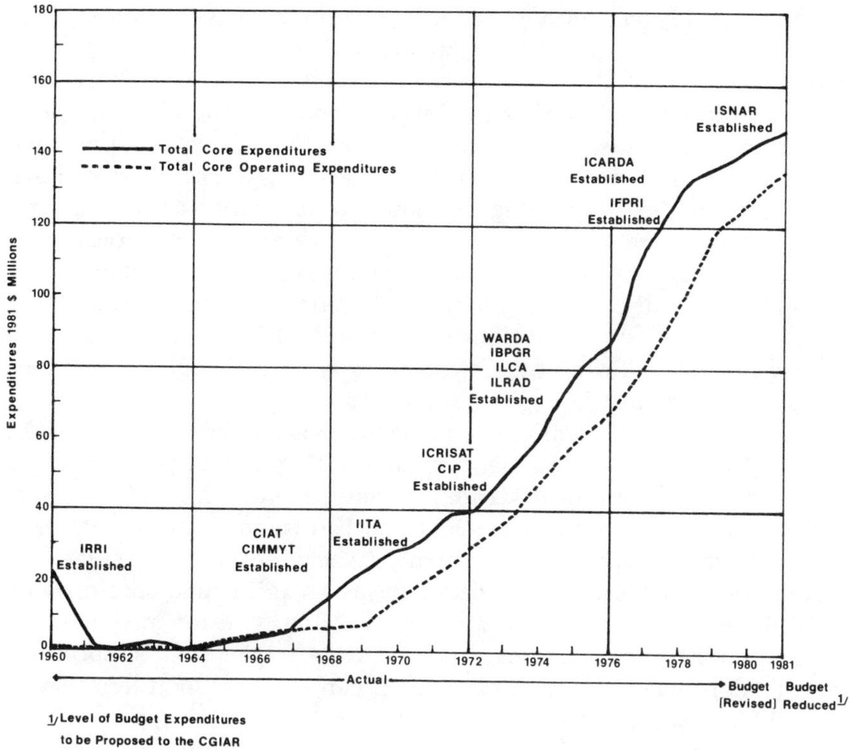

Figure 8.1 International Agricultural Research Centers' Annual Core Expenditures and Core Operating Expenditures, 1960-1980 (in terms of constant 1981 dollars). (From CGIAR Secretariat 1980, Washington, D.C.)

international agricultural research system deserve careful attention by the CGIAR system. These include relations with the research institutions of the developing countries, relations with the research institutions of developed countries, and financial and scientific program viability. The perspectives on these issues presented here draw on my experience as an IRRI staff member, a member of several institute review teams, and a member of the CGIAR's Technical Advisory Committee. I also draw on the 1976 and 1980 reports of the review team that the CGIAR commissioned to evaluate the work of the international agricultural research institutes.[6]

Relations with Research Systems in Developing Countries

By the late 1960s, many of the bilateral and multilateral aid agencies were recognizing serious shortcomings in their efforts to support the development of national agricultural research systems. Most national systems in the developing countries were unprepared to absorb effectively large amounts of financial, material, and professional assistance. The capacity for management and entrepreneurship among members of the newly trained scientific community was often underdeveloped. Many systems were plagued by cycles of development followed by erosion of capacity as budgetary priorities responded to change in political regimes.

Impatient staff members at aid agencies were often unaware of the history of their own national institutions. They had forgotten that the national agricultural research systems of the United Kingdom, Germany, the United States, and Japan had taken decades, not years, to acquire the research and training capacity needed to generate the knowledge and technology needed to sustain agricultural development. Furthermore, the political support available to many national and international aid agencies was often so fragile that support for institution building was difficult to sustain unless a short-term payoff could be visualized.

In addition to a sense of frustration with efforts to strengthen national research systems, staff members felt a growing conviction that the problem of meeting food requirements in the poor countries was urgent. The initial success of the IRRI's rice program and the CIMMYT's wheat program combined to create a feeling on the part of both aid donors and recipients that the international agricultural research institute, which could operate independently of the vagaries of the local political environment and could draw on the global agricultural science community for its staff, represented an effective instrument for the management of research resources and for the generation of agricultural technology.

Effective liaison between the international centers and national research systems was essential if individual countries were to make productive use of the knowledge and the technology generated at the international institutes. Efforts were made to establish collaborative research networks linking the international institutes with regional organizations supported by the FAO, the international banks, and bilateral aid agencies and with national research and training institutions. The linkages represented in the international wheat network are illustrated in Figure 8.2.[7]

By the mid-1970s, it had become increasingly clear that the productivity of the international agricultural research system was severely constrained by the limited capacity of many national research systems, and that the adaptation and dissemination of the knowledge and technology generated at the international institutes were dependent on the development of effective national systems. Robert E. Evenson's work has demonstrated that the ability to screen, borrow, and adapt scientific knowledge and technology required essentially the same capacity as is required to invent new technology.[8] Research capacity in the basic and supporting biological sciences is at least as important as capacity in applied science.

The outreach programs of the international institutes, even when working through networks such as the international wheat research network, the inter-Asian corn program, and others, did not have the capacity to take on the role of strengthening national systems. The regional commodity networks played an important role in enabling the institutes to conduct research, and to test materials and methods under diverse ecological conditions. But they could not assume a larger role without diverting effort from the institute's research programs. By the mid-1970s, only a few national systems—those of India, Brazil, and the Philippines, for example—had developed the managerial and professional capacity to absorb, transmit, and adapt effectively the knowledge and technology that were becoming available to them through the international institutes and from the research of the more developed countries.

The bilateral and multilateral assistance agencies had no alternative, therefore, but to place the strengthening of national research systems high on their agendas. Several steps have been taken by the CGIAR's system. The first direct effort was taken in 1973 when the TAC endorsed support for the coordinated trials program of the West African Rice Development Association (WARDA).[9]

The WARDA was established in 1968, with the support of the FAO and several bilateral donors, as a cooperative rice research and development program among eleven West African countries. The region is characterized by a great diversity of rice production systems—upland rice, irrigated rice, mangrove swamp rice, and deep-flooded and floating rice. The WARDA has a small professional capacity in its own staff, and it works in a region where national agricultural research is even weaker. Attempts were made to overcome some of these difficulties through cooperative programs with the IITA, the IRRI, and the Institut de Recherches Agronomie Tropicales et des Cultures Vivieres (IRAT). Efforts were also made to enhance available professional capacity by the program's decision to work in

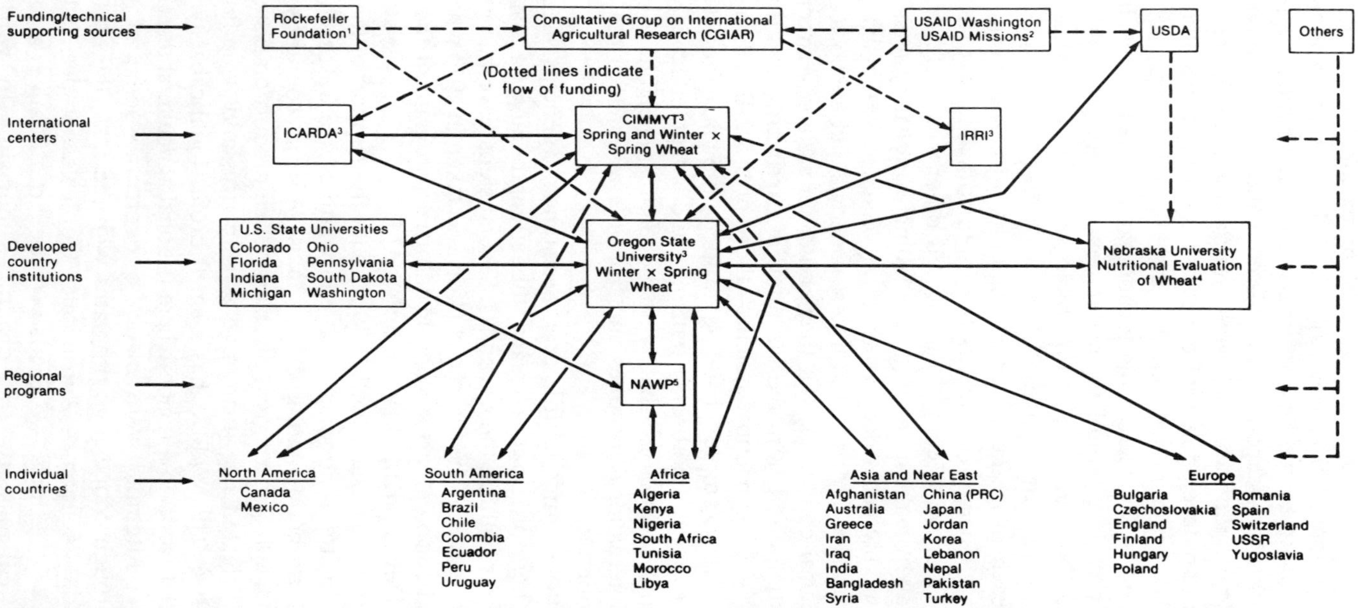

Notes: [1]The Rockefeller Foundation provides funds to Oregon State University for graduate training. [2]USAID missions contribute through bilateral support of individual country programs. [3]The links with individual countries take the form of information and materials exchange, the conduct of seminars and workshops, cooperative research projects, advisory services, and training. [4]USAID funding of nutritional work expired in late 1979. [5]North African Wheat Program. Source: Adapted from figure provided by Department of Crop Science, Oregon State University, Corvallis.

Figure 8.2 International Winter and Spring Wheat Research Network, 1979. (From Dana G. Dalrymple, *Development and Spread of Semi-dwarf Varieties of Wheat and Rice in the United States: An International Perspective*, Washington, D.C.: USDA, Agricultural Economics Report No. 455, June 1980, p. 133.)

collaboration with national programs. The WARDA's facilities and staff were located primarily at national research centers.

When I visited a number of WARDA and national facilities in 1978, in connection with the TAC's quinquennial review mission, I was impressed by the importance of the WARDA's linkage role. The effectiveness of researchers at the most remote stations, at Mopti in Mali, for example, was greatly enhanced through their access to the world collection of germ plasm and the communication network that made the methods and the results of rice research throughout the world available to them. Their participation in the WARDA's training activities and research conferences gave them a sense of sharing in a scientific effort that extended beyond the limited environment of their own stations.

A second step in the effort to strengthen national research systems has involved the development of a CGIAR-sponsored institution charged with supporting the development of national research systems. The Rockefeller Foundation played an important entrepreneurial role in this development. After a series of consultations with the leaders of national research systems, the International Agricultural Development Service (IADS) was established, with initial funding from the Rockefeller Foundation, to provide contract research management and development services to national research systems.

The initiative of the Rockefeller Foundation influenced the CGIAR to intensify its own deliberations. In 1977, the CGIAR organized a task force to explore the possibility of establishing an international service to strengthen national agricultural research within the CGIAR's systems.[10] These deliberations led to the establishment of the International Service for National Agricultural Research (ISNAR) in 1979. There had been some expectations that, in establishing the ISNAR, the CGIAR might absorb the IADS, much as it had assimilated the IRRI, CIMMYT, IITA, and CIAT in 1972. By 1979, however, the CGIAR had become somewhat sensitive about absorbing activities initiated prior to the CGIAR/TAC assessment and evaluation process.[11] Some European donors were also sensitive about the fact that staffing patterns at the institutes had not included sufficient numbers of European professionals. The FAO, one of the CGIAR's sponsors, expressed strong concern that the new service was infringing on an area of traditional FAO responsibility.

By 1984, the IADS had acquired substantial experience in managing projects, funded by agencies such as the U.S. Agency for International Development (USAID) and the World Bank, designed to strengthen national agricultural research systems. The ISNAR had acquired considerable expertise in diagnosing the problems that have inhibited the effectiveness of national research systems and in helping national agencies to reform and develop their own systems. But it was premature to evaluate the effectiveness of either effort. Clearly, however, strengthening national research systems is only partially, and perhaps only marginally, amenable to the efforts of the assistance agencies. At times, external funding agencies have inhibited the development of national systems because they have been insensitive to the problem of political and economic viability.[12] Unless national governments take the necessary steps to create viable national research systems—particularly those measures to establish hospitable and productive professional environments for indigenous professional capacity—the contributions of the bilateral and multilateral assistance agencies will have limited impact.

The international institutes have not always been the most effective instruments for reinforcing the emerging professional capacity in the national systems. Relationships have sometimes been competitive rather than collaborative. In efforts to establish their credentials and to provide the donors with evidence of productivity, the international institute staff and management have, in some cases, overemphasized their own contributions and failed to give appropriate credit to the work at national institutes. For example, the diffusion of the rice variety C-4, developed at the University of the Philippines College of Agriculture, did not receive the same support by the international system as did the diffusion of the varieties IR-5 and IR-8, developed at the IRRI. The IITA has viewed the WARDA as a competitor for donor resources rather than as a complementary resource.

Relations with the Research Institutions of Developed Countries

The initial years of the new international institutes were characterized by a tendency among the institutes to keep the universities and research institutions of developed countries at arm's length. The new institutes tended to view, and not without some justification, efforts by scientists from developed countries to establish collaborative relationships as efforts to exploit new sources for the funding of personal research interests that were only marginally complementary to the commodity-centered mission orientation of the new centers.

This relationship has changed over time. As the institutes have identified problems in which lack of knowledge in areas such as physiology, pathology, and other areas of science have become constraints on their ability to expand scientific and technical frontiers, they have taken steps to institutionalize their relationships with institutions from developed countries. Examples include the relationship among the CIMMYT and several Canadian institutions for work on triticale. The International Potato Center (CIP) has used contracts with institutions of developed countries more extensively than any of the other international centers. At the time of the 1977 quinquennial review mission, the CIP identified twelve such contracts with institutions from developed countries and seven with institutions from developing countries. In a number of cases, the CIP's contracts induced additional effort and expenditure on CIP-related problems by the contracting institution from the developed country.[13]

There are clear dangers in the growing relationships among the international centers and the centers of fundamental research in the developed countries. If the developing countries are to establish a viable base for self-sustained scientific effort, they must establish a capacity to work on the fundamental problems that are significant for tropical environments. In conversations with aid agency and university representatives from developed countries, exception is often taken to this view. The point is sometimes made that the scientific capacity already exists to conduct the relevant basic research and supporting research in the developed countries. What is the justification for duplicating expensive facilities? An appropriate response to this question is that the creativity of the scientist is strongly conditioned by the institutional and physical environment in which he or she works. Scientific imagination is as surely conditioned by the scientist's environment as the growth of the rice plant is conditioned by its environment.

In my view, a systematic effort to establish centers of fundamental research in the tropics, such as the International Center for Insect Physiology and Ecology (ICIPE) in Nairobi, should be encouraged. The international crop and livestock improvement centers should be encouraged to establish linkages with these institutions wherever they represent viable alternatives to institutions in developed countries. If such institutions cannot be funded within the CGIAR's system, donors from developed countries should seek another mechanism to ensure the development of centers of research and training in the supporting sciences that are related to the technology-oriented research of the international agricultural research system and of the national research systems.

Financial and Program Viability of the International System

The original four institutes (the IRRI, the CIMMYT, the CIAT, and the IITA) were initially funded jointly by the Ford Foundation and the Rockefeller Foundation. The CGIAR was organized to provide a broader resource base for the expanding system. Funding requirements have risen rapidly (see Figure 8.1), and funding problems have become more complex. The CGIAR provides a forum in which the international agricultural institutes set forth their program and funding needs and the donors make known their funding commitments. The commitments of the individual donors, however, are typically made to particular institutes rather than to the system as a whole. Although there is an exceptionally good record of funding continuity, institute directors (more recently, directors general) have had to devote an increasing percentage of their time to donor relationships in order to ensure the continuity and growth of funding. The fact that the funding is typically made available on an annual rather than a long-term basis puts particularly severe pressure on management to cultivate relationships with donors and to respond to their priorities.

During the early years of the system, the participation of the directors of these institutes in scientific leadership and personnel recruitment was an important factor in establishing the quality and productivity of the research program. By the mid-1970s, management had become more heavily involved in matters of financing and public relations, and the administration had become more highly bureaucratized. The external relationship responsibilities on the directors general are increasingly insulating them from program leadership. Most directors no longer have the professional capacity or the inclination to be vigorous leaders of the scientific programs of the institutes.

As the capacities of the national research systems in the developing countries improve, the relative contributions of the international agricultural research institute system to the generation of knowledge and technology will decline. One possible outcome of this process is that the institutes will lose the distinct leadership roles that they now possess. A viable model for the future of the institutes is an expansion of their role as centers for the conservation and diffusion of genetic resources and of scientific and technical information, relative to their role as producers of new knowledge and new technology. If they are careful to select staff members for their leadership capacities as well as for their scientific and professional competence, they will also be able to continue to play a strategic role in establishing research priorities.

There is, however, what might be considered a natural history of research institutes. A new institute that is able to bring together a team of leading scientists tends to go through a period of high productivity that often lasts a decade or longer. There is a tendency, after this initial period of creativity, for an institute to settle down to filling the gaps in the scientific literature and to fine-tuning incremental changes in technology.

The international institutes will need to guard against this tendency. There are numerous technical and institutional constraints that dampen the ability of farmers to achieve even present yield potentials economically. The world will continue to need a system of international institutes that will play a strategic role in the area of crop and livestock improvement. The political fragmentation that characterizes much of the world in which the international centers operate ensures that relatively few nations will become self-sufficient in their capacity for either training or research in the agricultural sciences.

If the international institutes develop a capacity to link the national systems into a carefully articulated international system, they will ensure their own continued viability. If they become viewed as being competitive with national research systems, they could fade away into mediocrity.

Future Policy and Program Directions

There are numerous pressures to broaden the responsibilities of international agricultural research centers: they are urged to become the primary centers for conservation of genetic resources, to expand their programs into research on cropping and farming systems, to develop research programs in pre- and postharvest mechanical technology, and to develop greater capacity in the basic services related to agriculture. They are also being pressured to place greater emphasis on small-farm technology and to become broad-based agricultural or rural development institutes rather than simple crop or livestock improvement centers.

Genetic Resources

Each institute has found it necessary to develop at least a working collection of genetic material for use in its crop improvement, technology development, and supporting science programs. Some institutes (such as the IRRI) have set out from the beginning to establish a world collection of genetic materials. The International Board for Plant Genetic Resources was established under the CGIAR's auspice's in 1974 to encourage and assist in the collection, preservation, and exchange of plant genetic materials.

Upon its inception, the International Board for Plant Genetic Resources (IBPGR) faced a situation in which the motives and aims of the several available collections were very different. Many useful collections, kept by individual plant breeders, included only the cultivars that had proven to be most useful to a particular breeder's program and did not represent the full range of heritable variation. Such collections were often dispersed or lost because of a lack of funds or the retirement of the leader of a breeding program.

At the other end of the spectrum were great conservation collections containing a substantial part of the genetic variation of a wide range of species, such as those of the National Seed Storage Laboratory at Fort Collins, Colorado, in the United States and the N.I. Vavilov Institute in the Soviet Union. Some national centers have functioned more as museums than as centers of plant breeding. Among the international institutes, the IITA in Nigeria, the IRRI in the Philippines, and the International Crop Research Institute for the Semi-Arid Tropics (ICRISAT) in India had assembled large genetic resource collections that were intimately related to the breeding programs of the institutes. Most collections consist exclusively of seed materials. The problems of storing genetic material for crops that are normally propagated with vegetative materials (such as cassava, potatoes, and yams) remained unresolved.[14]

It is urgent that decisions with respect to the appropriate locations for the facilities and programs for collection, maintenance, and use of genetic resources be made. Issues regarding the most effective way of linking the various collections, establishing exploration and collection priorities, and making materials available to international and national breeding programs are not easily resolved. Albert Moseman, in a personal correspondence (May 1977), has argued convincingly that, if genetic resource centers are to make productive contributions to the development of biological technology, they must be institutionalized in locations where strong plant breeding and genetics research programs are being conducted.

In my judgment, a strong argument can be made for having each of the major international crop institutes evolve into a world center for the conservation and evaluation of genetic materials. Even if contributions to knowledge and technology at the international centers decline relative to the contributions of the research institutions of developing countries, the national agricultural research systems will still need a reliable source for materials and information on genetic resources and potentials.

The budgetary requirements for a serious commitment to genetic conservation and use are likely to be larger than many of the donors to the international system realize or the directors general of the international institutes are willing to admit. The CGIAR donors and the institute management will face continuous pressure for diversion of financial resources from support of the genetic material archives to use in technology development research unless the genetic material collections are seen as contributing in a very direct way to technology development. The value of the world collections to technology development will be most apparent if the collections are organized and managed as part of projects that utilize genetic resources.

Farming and Cropping Systems Research

The initial programs of the international centers focused primarily on crop improvement and advances in the technology of crop production. Some institutes were given responsibility for research on a single crop. Even those institutes that had responsibilities for more than one crop often functioned as a set of loosely linked single-commodity institutes held together by the economies of scale in administration and related institutional infrastructure (motor pool and library, for example).

Farming or cropping systems research was included in the mandates of several of the newer institutes (the CIAT, the IITA, and the ICRISAT, for example) at the time the institutes were established. Several of the other institutes have also established farming or cropping systems programs. Although the CIAT has discontinued its farming systems program, it has continued to work on cropping systems within the framework of its individual commodity research program. At several of the institutes, the cropping or farming systems programs have evolved as an important instrument to achieve more effective integration of disciplinary research.

The initiation of farming and cropping systems research has clearly been one of the exciting contributions of the international agricultural research system. In most national programs, research efforts have been fragmented along disciplinary lines. Even scientists from the two integrating fields of agronomy and farm management are often confronted with institutional barriers to effective collaboration.

The farming and cropping systems initiatives at the international institutes have, therefore, generated a great deal of intellectual excitement among scientists and agricultural program managers who were looking for a way to integrate the knowledge emerging from disciplinary and commodity research programs with farming practice.

The issues surrounding the appropriate role of farming and cropping systems programs within the international institutes have, however, been difficult to resolve.[15] Even the terminology that is appropriate to categorize such programs has been disputed. The issue of the appropriate role of farming and cropping systems research at the international agricultural research institutes must be evaluated, but not on the basic merits of the cropping and farming systems methodology, but in terms of the specific role that the international institutes can reasonably be expected to play in the development and dissemination of knowledge, materials, and technology.

A modest program of farming or cropping systems research is clearly essential even if a research institute takes as its primary goal the limited objective of developing improved plant materials—and other technical components, such as plant protection technologies—for a single commodity. Research directors and program leaders must have accurate information about the technical and environmental constraints influencing the productivity of the farming systems that include the commodities on which the institutes are conducting research. Knowledge about the cropping and farming systems employed by farmers in the major agroclimatic regions in which the institutes work represents an important part of the knowledge needed to allocate resources among institute research activities.

At some institutes, the cropping systems programs have broader objectives. One stated objective is to develop the methodology of farming and cropping systems research and to transfer such methodology to national systems. Another stated objective is to develop more productive cropping systems that can be extended to farmers.

The first objective has considerable merit. There are areas in which radical departures from traditional practice may be required for the evolution of viable intensive farming systems. The development of so-called prototype farming systems for such environments may, in some cases, be an appropriate objective of an international institute. The maize-based zero tillage system at the IITA is sometimes cited as an example of a prototype system that is approaching the stage at which further development and testing by national or provincial systems is warranted.

The second objective is not, in my judgment, a credible objective for cropping systems research at the international institutes. The fine-tuning of cropping systems varies according to location. Such fine-tuning is more appropriately a function of the local (state and prefectural) stations of national systems than of the international institutes.[16] Although modest productivity gains may be made as a result of cropping systems research at the international centers, it is unlikely that these gains will approximate those that can be realized from the resources devoted to the development of improved components for cropping systems.

Pre- and Postharvest Mechanical Technology

The allocation of resources for the development of pre- and postharvest mechanical technology has been an issue that has continuously faced the management of the individual institutes, and the CGIAR and its Technical Advisory Committee.

A small-machinery development program was initiated at the IRRI in the mid-1960s with support from USAID. Limited work on small machinery has also been initiated at the ICRISAT and the IITA. The program at the IRRI has included research and development work on irrigation, land preparation, and weed and pest control, as well as on harvesting, threshing, and drying equipment. The IRRI's program has had imaginative engineering leadership and has developed an effective outreach program with manufacturers of agricultural machinery in the Philippines. After two decades of effort, however, the machinery development program still must measure the impact of its machine designs in thousands of units manufactured and distributed in contrast to the work on genetic improvement, which measures its output in millions of hectares planted to new varieties or millions of metric tons of additional rice population. The impact of the machinery development programs at the IITA and the ICRISAT are even less impressive.[17]

Considerable discussion of the role of research on postharvest technology at the international institutes has ensued. The establishment of a postharvest research and training institute within the CGIAR's system has been suggested. Although neither of the options has been implemented, several of the CGIAR's donors committed themselves in 1976 to supporting an information and advisory service on postharvest research to be organized under the auspices of the Southeast Asian Regional Center for Graduate Study and Research in Agriculture (SEARCA) in the Philippines. The gains in postharvest technology, though potentially large, have thus far been achieved at a relatively high cost or else have generated relatively low returns.[18]. Experience suggests that both the expertise needed to conduct research and the articulation of research with technology development and dissemination are most effective when postharvest technology development is conducted in the private rather than in the public sector.

The evidence with respect to the assumed complementarity between tillage equipment and cropping intensity that has provided a rationale for the small-machinery development programs is less than compelling. It is hard to find evidence that the private sector lacks incentives for the development of pre- and postharvest technology when economic conditions encourage a demand for such equipment. The extremely rapid diffusion of the power tiller and mechanical tillage equipment in Taiwan since the mid-1960s, and rice-milling equipment on Java since the early 1970s, seems to suggest that in most areas demand constraints have been more serious than supply constraints in the diffusion of mechanical technology.

The reasons for the greater effectiveness of public sector (or philanthropic) research activity in biological and chemical technology compared to that in mechanical and engineering technology are fairly straightforward. The invention of new biological and chemical technology for agriculture typically requires high-level scientific skills in genetics, physiology, pathology, chemistry, statistics, and related disciplines. The design of new machine technology in agriculture depends to a much greater extent on mechanical skill and ingenuity, and on farming experience and practice. Furthermore, the relevant skills in mechanical technology are pervasive in developed countries and are widespread even in developing countries except for the most backward countries of the "fifth world."

In my judgment, work on equipment design at the international agricultural research institutes should be limited to work on prototype equipment needed for radically new production technologies or farming systems that have not previously been available to farmers. It would be inadvisable for the institutes that are members of the CGIAR system to expand their machinery design or postharvest technology programs without more careful analysis of the technical and economic potential of such programs than has been done in the past.

Small-Farm Technology

The issues of differential access of large and small farms to new crop varieties, and to crop production technologies generated by the international agricultural research institutes, have been the subject of interdisciplinary conflict almost from the time the new wheat and rice varieties appeared on the scene. The view was expressed that the new institutes represented "big science for big farmers." The small farmer was viewed as an actual or potential victim rather than as a beneficiary of the new technology.

Much of this criticism was based on the crudest sort of casual empiricism. Some of it appeared to reflect little more than interdisciplinary pique or aggression. The Green Revolution had shifted attention from the early architects of development—planners, economists, and social scientists—to geneticists, plant breeders, and agronomists. Critics, who welcomed the Green Revolution's technology in Cuba and China, viewed the introduction of the high-yielding varieties in nonsocialist economies as raising the cost of radical change by channeling new income to the middle- and upper-level peasantry.[19] Many sponsors of the CGIAR's system reacted to these criticisms by calling for a redirection of scientific and technical effort at the institutes. Sometimes the alteration was cast in terms of producing labor-intensive or worker-friendly technologies. Sometimes it was cast in terms of technology for small farmers.

What has not been generally appreciated in this argument is the fact that in labor-intensive, low-wage economies, simply focusing technical effort in the direction that would have the highest payoff will bias the generation of technology in the direction desired by the critics. Both economic logic and empirical evidence support the view that the generation of a technology designed to use relatively scarce and expensive inputs—a capital-intensive technology in a labor-surplus economy, for example—would be less profitable and diffuse more slowly than a technology designed to absorb more abundant and less expensive labor inputs.[20]

But the development of appropriate technologies, in the above sense, is precisely what most of the institutes were doing. The emphasis on new biological technology rather than mechanical technology resulted in the development of technologies that were essentially scale neutral. The decision at the CIAT to deemphasize the beef program and to place increased emphasis on beans was a decision consistent with the factor endowments in the poorer countries and regions of Latin America—and consistent as well with developing technologies for small farmers.

When the hard data began to come in and to replace the casual observations that had fueled the early arguments, it became apparent that the technology of the Green Revolution was being adopted at least as rapidly by small farmers as by large farmers in the areas where the new technology had been adapted to the agroclimatic conditions. This has been shown rather dramatically in a series of studies conducted by the IRRI and cooperating national institutions on thirty villages in six countries in Asia. On the average, small farmers have adopted the new high-yielding technology faster than large farmers (see Figure 8.3). One reason this has occurred is that the technology was suited to those crops produced by small farmers and those regions characterized by small farms.

Once the rhetoric of ideological debate is left behind, two difficult issues remain. The first is whether technologies can be invented that confer a larger advantage on small farms than on large farms. A small-farm bias may be introduced through selection of the commodities mix on which to conduct research. Research on beans in Latin America and dairy production in India are examples. The farming systems research program of the Centro Agronómico Tropical de Investigación y Enseñanza (CATIE) in Central America is another example. Once the geographic and commodity focus of a research effort has been made, however, it is very difficult to find many examples of technology that can be biased strongly toward small farmers. It seems rather clear that institutional changes, in land-tenure relations for example, have a

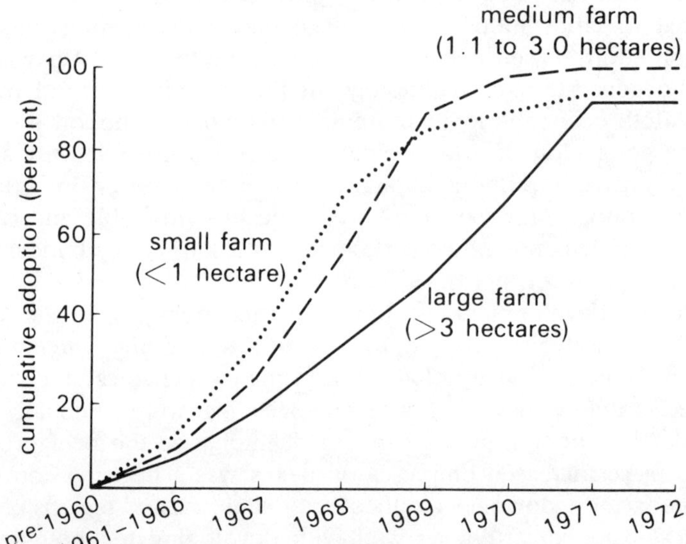

Figure 8.3 Cumulative Percentage of Farms in Three Size Classes Adapting Modern Varieties in 30 Villages in Asia. (Adapted from Randolph Barker and Robert W. Herdt, "Equity Implications of Technology Changes," in *Interpretive Analysis of Selected Papers from Changes in Rice Farming in Selected Areas of Asia,* Los Banos, Laguna, Philippines: International Rice Research Institute, 1978, p. 91.)

potentially much more powerful influence on the distribution of gains from technology than do attempts to go beyond the objective of designing technology that will be profitable for the farm structure and resource endowments that actually exist.

The second major issue is whether productive technologies can be developed for regions where the soils are poor, the slopes are steep, or the rainfall is low. Can productive low-input technologies be developed? Or must the ultimate solution be the reduction of cropping pressure in marginal areas combined with more intensive use of more favorable areas? All of the technical answers to this question are not yet in. And the institutional answers will depend on the pace of development in the nonfarm sectors of the economies of which the more fragile environments are components.

Agricultural Development Centers

There is also continuous concern that the institutes should develop an analytical capacity and a flow of research results designed to influence agricultural and rural development policies. This pressure has risen with the perception that institutional constraints—such as the intervention into input and product markets, the performance of extension, credit, and market institutions, and the capacity for investment and management of physical infrastructure such as irrigation systems—represent increasingly serious constraints on the realization of the productivity gains that the new biological technology makes possible.

By the early 1960s, it had become quite clear that low returns were being realized from investments directed toward institutional reform and institutional innovation. The response by aid donors was a redirection of institution-building efforts to focus on creating the research capacity needed to produce a continuous stream of more productive biological technology. This effort has been highly successful.

It seems clear that it is now time again to redirect the capacity for institutional innovation toward enabling rural people in the developing countries to realize the potential gains from the higher productivity—higher output per hectare and per worker—that is becoming available. This shift in the demand for institutional innovation is creating a derived demand for new knowledge in the rural social sciences. How should the international crop and livestock research institutes respond to this perceived growing demand for the social science knowledge that is needed as an input into institutional innovation?

Clearly, a strong social science research capacity can represent a highly productive use of institute or center resources even if the primary objective of the center remains simply the invention or design of improved components for cropping systems. The economics staff of the IRRI, for example, is able to provide the research director and the program leaders with relatively precise information regarding the value of the potential productivity gains that can be realized by alternative efforts to release yield constraints under farm conditions. The staff is also able to identify with some precision the extent to which productivity gains depend on investment or management decisions that lie outside the scope of the IRRI's research program and that are amenable only to policy intervention or institutional innovations over which the IRRI has little influence.[21]

In 1973 and 1974, the CGIAR and its Technical Advisory Committee initiated a series of consultations about the role of social science research at the international institutes and in the international system. One result was a modest strengthening of social science research capacity at several of the institutes. Another result was the recommendation that the International Food Policy Research Institute (IFPRI) be established to conduct research on agricultural production and price and trade policy issues. Initially, the recommendation was not accepted by the CGIAR. Some donors felt that an institution designed to analyze the burden of agricultural and food policy on agricultural development in the developing countries would not be able to avoid turning its attention to the effects of policies of the developed countries, particularly the policy of the European Common Market on the agricultural export potential of the developing countries. The IFPRI was, however, established in 1975 in Washington, D.C., by a group of CGIAR donors (the Ford Foundation, the International Development Research Center [IDRC], and the Rockefeller Foundation), and in 1979 it was finally incorporated into the international agricultural research system.

My present perspective is that each institute needs a social science research staff of at least the size and capacity that was established at the IRRI and the ICRISAT by the late 1970s if the directors and program leaders are to be in a position to allocate effectively the research resources available to them. However, even at the IRRI and the ICRISAT there would be a substantial payoff to increasing the staff's capacity in anthropology and sociology relative to economics. Anthropologists, in particular, have demonstrated a capacity to understand the dynamics of technology choice and impact at the household and village level that is highly complementary to both agronomic and economic research.

The internal dynamics of a staff's interrelationships and professional balance will, however, make it difficult for directors to allocate a very significant share of institute resources to social science research on more macroeconomic agricultural and rural development issues. The international system should seek ways to strengthen graduate training capacity in the rural social sciences and research capacity at the central research or ministry level within the national training and research institutions of developing countries.

A Perspective on the Future

A major obstacle to the future development of the international system is the spectacular contribution to productivity growth in rice and wheat resulting from research conducted at the IRRI and the CIMMYT. Both institutes were able to build on a large backlog of relevant knowledge and to work on a major gap between the yields being obtained on the same crops under tropical and temperate conditions.

Even the most conservative estimates of the result of the research conducted at the IRRI suggest that, by crop year 1974/75, the supply of rice in all developing countries was approximately 12 percent higher than it would have been if the same resources had been devoted to the production of rice using only the traditional rice varieties available prior to the mid-1960s.[22] The impact of the CIMMYT's research on wheat production has probably been even greater.

One result of the very high rates of return to the initial investments in research at the IRRI and the CIMMYT was to create expectations about the performance of the other institutes in the CGIAR's system that have been difficult to realize. Other centers are working on crops and farming systems for which the research base at the time their programs were established was much less adequate than for rice and wheat. It seems reasonable to anticipate much more modest returns on investment in the CGIAR system in the future than in the past. In the future, the institute's accomplishments will probably be described in terms of incremental gains rather than in terms of revolutionary breakthroughs. The returns from such research are, however, likely to be significantly higher than the returns that can be realized on most of the other investments in agricultural or rural development that are available to the CGIAR's donors.

A second result of the early success in crop technology is the credibility generated by institutes in areas in which they have little capacity. This has at times led to an increase in efforts by donor agencies to channel bilateral technical assistance funds through one or another of the international assistance programs. The financial pressures operating on the directors of individual institutes are difficult for them to resist. It will be necessary for the CGIAR's centers to avoid taking on projects that divert the limited capacity for scientific management to low-productivity activities. In the future, the system will have to develop greater institutional capacity than it has in the past to discontinue unproductive activities and to avoid adding activities that are marginal for its mission.

The international agricultural research system should be viewed as a permanent feature of the global agricultural development infrastructure. Its significance as a source of new knowledge and new technology is not limited to its exploitation of the potential for low-cost productivity gains that were available during the third quarter of the twentieth century. This is not to suggest that the program of each of the present institutes should continue indefinitely. The charter of each institute has a dissolution clause. It is quite possible that individual centers that fail to meet CGIAR expectations with respect to technology development should be phased out.

I do have several major concerns about the future of the international agricultural research system. One is that the managerial structure of the system might, in the future, become overly bureaucratic. It is possible that the informal funding arrangements and decentralized managerial structure may protect the system from the evolution of a burdensome administrative structure either at the level of the CGIAR or the TAC Secretariat. One student of science policy, when confronted with the organizational and financial structure of the system, insisted that "it can't exist." Excessive centralization of management and planning could dampen incentives for scientific entrepreneurship and productivity.

A second concern I have is with the evolution of a steering mechanism that will ensure a continuing focus on the most significant research problems. As each new institute was established, priorities were set through an extended series of consultations at both scientific and policy levels. As a research institute matures, it becomes increasingly difficult to redirect its efforts. In the U.S. and Japanese systems the tension between the national and state (prefectural) systems, and the involvement of clientele organizations, had acted to induce an efficient portfolio of projects. In the international system, the clientele—the developing countries—is only beginning to become involved in a significant way in the financial support of the system. This limited financial involvement weakens the weight given to the perspectives of the developing countries in research policy discussions at the individual institute governing boards and at the CGIAR level.

Third, I continue to be concerned about how slowly national research systems in the developing world are acquiring the capacity to make effective use of the support that has become available as a result of the organization of the international system. The number of staff devoted to agricultural research in the developing countries is small.[23] Many research institutions in developing countries are poorly managed. Consistent political support for the funding of agricultural research has been difficult to institutionalize. The effectiveness of the international system depends on the development of strong national systems.

Finally, I am concerned about the capacity of the international institute to attract and hold world-class scientists. One source of this concern is the widening gap between the staff requirements of the international institute and the direction of training at the doctoral and postdoctoral levels in U.S. universities. In the past, a relatively large share of the staff of the international institutes, both from the developing and developed countries, has been trained at U.S. land-grant universities. As the emphasis in U.S. graduate and postdoctoral training continues to shift toward greater emphasis on basic science and away from applied science and technology development, U.S. training may no longer be as compatible with the needs of the international centers or of national research systems in developing countries as in the past. A second source of concern is that science careers, including careers in agricultural science, appear to be less attractive to the most qualified U.S. undergraduates than in the past. This appears to be, at least in part, one of the reasons for the declining share of domestic graduate student enrollment in colleges of agriculture.

Notes

1. This chapter is a revised version of "Toward a Global Agricultural Research System: A Personal View" Research Policy 15 (December 1986), pp. 307-327. Earlier versions of Research Policy (Minneapolis: University of Minnesota Press, 1980), pp. 116-160; in Vernon W. Ruttan, "The Role of International Agricultural Research Institute as a Source of Agricultural Development" (1978), pp. 298-308, and in Hueg and Gannar (1978), pp. 124-138. Many of the issues discussed in this chapter are also discussed in Per Pinstrup-Andersen (1982), and in Wortman and Cummings (1978), pp. 113-143. I have benefited from comments made on an earlier draft by Per Pinstrup-Andersen, Randolph Barker, Hans Binswanger, Nyle Brady, John Coulter, Ralph W. Cummings, Dana Dalrymple, Jan Flora, James Goering, Robert W. Herdt, Richard DeMuth, Haldore Hanson, Wiliam

Gamble, Max Langham, Kenneth Menz, John Nickel, Peter Oram, Edward Oyer, Albert Moseman, Ralph Retzlaff, and Grant Scobie.

2. For a history of the establishment and early development of the FAO, see Hambidge, (1955).

3. For a more complete background on the establishment of the IRRI and the CIMMYT, see Stakman, Bradfield, and Mangelsdorf (1967). For a discussion of the organization of the international agricultural research system, see Crawford (1977), pp. 281-294.

4. Schultz (1964).

5. Ruttan (1969), p. 226.

6. McCalla, et al. (1976). A second review of the institute system was initiated in 1980 under the direction of Michael H. Arnold (1981), pp. 15-21. For more detailed information the reader will find the quinquennial reviews of individual institute programs, available from the CGIAR secretariat, of value.

7. For greater detail, see the material prepared by Dalrymple, (1977), pp. 44-133.

8. Evenson, (1977), pp. 209-236.

9. For a more complete review of the WARDA program, see Technical Advisory Committee (1979). In retrospect, the 1979 TAC report appears unduly optimistic about the WARDA experience and potential.

10. See Consultative Group for International Agricultural Research (1978).

11. After a good deal of deliberation, the CGIAR had earlier decided not to provide funding, under the CGIAR's auspices, for the Asian Vegetable Research and Development Center (AVRDC), the International Fertilizer Development Center (IFDC), or the International Center for Insect Physiology and Ecology (ICIPE). The reasons were somewhat different in each case. In the case of the AVRDC, the uncertain political status of Taiwan was an issue. In the case of the IFDC, the reluctance to open the door to the direct support of institutes in developed countries was of concern. In the case of the ICIPE, a reluctance to fund what some donors viewed as basic research seemed to be involved.

12. Ruttan, (1983), pp. 405-417.

13. See Technical Advisory Committee (1977), pp. 85-86.

14. See International Board for Plant Genetic Resources (1975, 1980).

15. Technical Advisory Committee (1978). See also Binswanger, Krantz, and Virmani (1976).

16. This view is generally consistent with, but somewhat less optimistic than, the perspective presented in Gilbert, Norman, and

Winch (1980).

17. The problem of the limited effectiveness of public sector machinery development programs remains something of a puzzle. There is a large body of experience that suggests that the private sector, even in developing countries, has been much more effective in the development and diffusion of mechanical technology than the public sector. There is, for example, a vigorous small-scale farm machinery and equipment industry in Thailand that has been innovative in developing new designs suited to the physical and economic environment in which the machinery and equipment is to be used. There is also a good deal of confusion about the "need" for mechanization. It is frequently asserted that when there is a short turn-around time between crops, mechanization is "necessary" to permit double or triple cropping. But in Taiwan, a multiple-cropping index well above 2 was achieved before significant mechanization. The issue of termlines should be cast in economic rather than physical terms. Where labor is abundant (or cheap) relative to land, lack of mechanization has not been a restraint on cropping intensity. Where labor is scarce (or expensive), mechanization can make an important contribution to breaking the labor constraint and to enhancing output.

18. For a more positive view, see National Research Council, (1978).

19. For a more detailed discussion of the arguments, see Binswanger and Ruttan, (1978), pp. 381-408; and Hayami and Ruttan, (1985), chap. 11.

20. See Scobie, (1979), p. 25.

21. See, for example, International Rice Research Institute, (1979). See also Sanders, (1980).

22. Evenson, Flores, and Hayami, (1978), pp. 243-265. The authors pointed out that there are several ways of estimating the advantage of the high-yielding varieties. They estimate that under farm conditions the yield advantage of the new high-yielding varieties over traditional varieties was in the order of 6 percent in South Asia and 20 percent in Southeast Asia during the early 1970s. The cost advantage of producing a ton of rice using high-yielding varieties was about 25 percent in South Asia and 15 percent in Southeast Asia (p. 13). See also the results presented in Evenson's Table 11. At the time this paper was being completed the CGIAR was initiating a comprehensive study to evaluate the economic impact of INRC technology.

23. A report by the World Bank, (1980), noted that in the entire developing world the number of agricultural research staff members (about 23,000) is hardly more than the number in Japan alone.

References

Arnold, Michael H. Second Review of the CGIAR, *Report of the CGIAR Review Committee*. Washington, D.C.: CGIAR Secretariat, World Bank, November 1981.

Binswanger, Hans P., B. A. Krantz, and S. M. Virmani. *The Role of the International Crops Research Institute for the Semi-Arid Tropics in Farming Systems Research*. Hyderabad, India: ICRISAT, April 1976.

Binswanger, Hans P., and Vernon W. Ruttan. *Induced Innovation: Technology, Institutions and Development.* Baltimore: Johns Hopkins University Press, 1978.

Consultative Group for International Agricultural Research (CGIAR). *Report of the Task Force on International Assistance for Strengthening National Agricultural Research*. Washington, D.C.: CGIAR Secretariat, World Bank, August 1978.

Crawford, J. G. "Development of the International Agricultural Research System." In *Resource Allocation and Productivity in National and International Agricultural Research*, Thomas M. Arndt, Dana G. Dalrymple, and Vernon W. Ruttan, eds. Minneapolis: University of Minnesota Press, 1977, 281-294.

Dalrymple, Dana. "Global Agricultural Research Organization." In National Research Council, *World Food and Nutrition Study,* Supporting Papers 5. Washington, D.C.: National Academy of Sciences, 1977.

Evenson, Robert E. "Cycles in Research Productivity in Sugarcane, Wheat and Rice." In *Resource Allocation and Productivity in National and International Agricultural Research*, Thomas M. Arndt, Dana G. Dalrymple, and Vernon W. Ruttan, eds. Minneapolis: University of Minnesota Press, 1977.

Evenson, Robert E., Piedad M. Flores, and Yujiro Hayami. "Costs and Returns to Rice Research." In *Economic Consequences of the New Rice Technology*. Los Banos, Philippines: International Rice Research Institute, 1978.

Gilbert, E. H., D. W. Norman, and F. E. Winch. *Farming Systems Research: A Critical Appraisal.* Rural Development Paper No. 6. East Lansing: Michigan State University, May 1980.

Hambidge, Gove. *The Story of FAO.* New York: Van Nostrand, 1955.

Hayami, Yujiro, and Vernon W. Ruttan. *Agricultural Development: An International Perspective,* rev. ed. Baltimore: Johns Hopkins University Press, 1985.

Hueg, William F., Jr., and Craig A. Gannar, eds. *Transforming Knowledge into Food in a Worldwide Context.* Minneapolis: Miller Publishing, 1978.

International Board for Plant Genetic Resources. *The Conservation of Crop Genetic Resources.* Rome: IBPGR, FAO, 1975.

——. *Annual Report, 1979.* Rome: IBPGR, FAO, May 1980.

International Rice Research Institute. *Farm Level Constraints to High Rice Yields in Asia.* Los Banos, Philippines: IRRI, 1979.

McCalla, Alex, Ewert Aberg, James McWilliams, and Arthur Mosher. *CGIAR Review Committee, Final Report.* Washington, D.C.: CGIAR Secretariat, World Bank, October 1976.

National Research Council. *Post-harvest Food Losses in Developing Countries.* Washington, D.C.: National Academy of Sciences, 1978.

Pinstrup-Andersen, Per. *Agricultural Research and Technology in Economic Development.* London and New York: Longman Group, 1982.

Ruttan, Vernon W. "The International Institute Approach." In *Agents of Change: Professionals in Developing Economies,* Guy Benveniste and Warren F. Ilchman, eds. New York: Praeger, 1969.

——. "The Role of the International Agricultural Research Institute as a Source of Agricultural Development." *Agricultural Administration* 5 (1978): 298-308.

——. *Agricultural Research Policy.* Minneapolis: University of Minnesota Press, 1980.

——. "Reforming the Global Agricultural Support System." In *Issues in Third World Development,* Kenneth C. Nobe and Rajan K. Sampath, eds. Boulder, Colo.: Westview Press, 1983.

Sanders, John H. "After the Dwarf Varieties: Crop Program Strategies and the Role of the Economist." Mimeographed. Cali, Colombia: International Center for Tropical Agriculture, September 1980.

Schultz, Theodore W. *Transforming Traditional Agriculture.* New Haven: Yale University Press, 1964.

Scobie, Grant M. "Investment in International Agricultural Research: Some Economic Dimensions." Staff Working Paper No. 361, World Bank, Washington, D.C., 1979.

Stakman, E.C., Richard Bradfield, and Paul C. Mangelsdorf. *Campaigns against Hunger.* Cambridge: Harvard University Press, 1967.

Technical Advisory Committee (TAC). *Report of the TAC Quinquennial Review Mission to the International Potato Centre.* Rome: TAC Secretariat, FAO, 1977.

——. *The Review of Farming Systems Research of the International Agricultural Research Centers: CIAT, IITA, ICRISAT, and IRRI.* Rome: TAC Secretariat, FAO, April 1978.

——. *Report of the TAC Quinquennial Review Mission to the West African Rice Development Association.* Rome: TAC Secretariat, FAO, 1979.

World Bank. *Agricultural Research Systems-Sector Policy Paper.* Washington, D.C.: World Bank, 1980.

Wortman, Sterling, and Ralph W. Cummings. *To Feed This World: The Challenge and the Strategy.* Baltimore: Johns Hopkins University Press, 1978.

CHAPTER 9

Farming Systems Research and Extension

ROBERT K. WAUGH
PETER E. HILDEBRAND
CHRIS O. ANDREW

The term "farming systems research" has been applied over the last decade to several agricultural development activities around the world: Colombia (Zandstra et al. 1979), Senegal (Shaner et al. 1982), Philippines (Harwood 1979), Mexico (CIMMYT 1974), Guatemala (Waugh 1975), Nigeria (Abalu 1983), Botswana (Norman 1983), and Panama (Tarte 1983), and elsewhere (Whyte 1981; Shaner et al. 1982). From these projects came ideas that agricultural economists especially were active in expressing (Gilbert et al. 1980). The roots of a new approach had sprouted and many people foresaw valuable features in what they observed.

The focus given to these development activities grew out of technical assistance programs in the Third World countries after World War II when domestic governments, international financial organizations, and private foundations sought to increase food production and improve the well-being of the world's populations living in stressed areas.

Following World War II the research and extension sponsored by international agencies in the developing countries had been largely patterned from U.S. domestic practice. The United States was the leader in supporting agricultural development in the developing countries, and at first emphasized transfer by promoting the U.S. concept of extension (Rice 1974). When it was found that most technologies were "location specific" and that adaptation must necessarily precede adoption, adaptive (frequently called applied) research and institution building were initiated to develop the local capability for carrying out the adaptations. This adaptive research phase had some successes. Several national research/extension programs were set up, the International Agricultural Research Centers

(IARCs) were established, and the Green Revolution occurred. Total food production began to increase more rapidly in developing countries than in the developed world. Per capita food production in the developing countries, however, lagged behind the developed nations (Barr 1981; Waugh 1982). Food production increased in only some countries, and even though applied agricultural research functioned concurrently with major community development projects, rural poverty was pervasive in many countries. Much of the poverty was found precisely where much of the basic food was produced—on the small, limited-resource farms.

Agricultural researchers whose work was directed toward the development of biological technologies (plant, animal, and soils researchers) viewed their contributions as a thrust to increase total food output rather than as a prime instrument for development and equitable distribution. They did not view small, limited-resource farmers as principal clients for their objectives. The plight of the small farmer was such a major dimension, however, that it was frequently discussed among agricultural scientists and community development project analysts. Scientists began to look at the cropping systems of small, limited-resource farmers. Bradfield's work at the International Rice Research Institute (IRRI) promoted the idea that mixed cropping has some advantages under the traditional systems of farmers in the Third World. When mixed-cropping systems that were developed on experiment stations were introduced to farmers, however, they were not readily accepted. This led to further shifts in the thinking of scientists. Harwood, who followed Bradfield at IRRI, emphasized the need for change, stating "to accomplish the development of a greater number of the world's small farms, shifts in emphasis must be made in our thinking, in our technological research and in our communications with farmers" (1979: xiii). By the end of the 1960s several approaches had been tried; new ones were to be tried, and are still being tried.

"Community development" projects were a major focus of developmental assistance during the 1950s and early 1960s (Ruttan 1984). Integrated rural development projects of the late 1960s and early 1970s were their sequel. The latter recognized that rural development involved a large number of interrelated activities, the coordination of which required developing a systems approach. It meant creating an awareness that the modification of one component of a system could materially affect another. A second perception of the integrated rural development approach was that more productive technology did not ensure equitable sharing of income—a major deficiency of the Green Revolution and commodity-oriented production campaigns. Since the integrated agricultural research/extension systems approach is still the approach used today, it is only logical that we identify some of its first project roots. Ruttan (1984) cites the Puebla Project of CIMMYT and the Rockefeller Foundation as an example of an integrated project that focused on agricultural technology and the conditions under which technology would benefit the small farmers of the area. While these activities have had important common threads, the overall concepts, strategies, and methodologies differed somewhat. Some of the common threads that bound together the early efforts of the farming systems approach and that survive today are:

1. A concern with limited-resource farmers and their families who have not adapted high-yielding technologies to the agricultural and socioeconomic environments of their farms
2. Recognition that an understanding of the farmers' situation is essential for developing technology appropriate to the conditions of limited-resource farmers
3. The use of scientists from more than one discipline working together as a team in order to understand the farm as an entire system rather than studying isolated components within the system
4. Understanding that new technologies and potential solutions to problems should be evaluated within the systems of production practiced by the farmers, under their conditions, and with farmer participation

Information and Technology for Policymakers and Farmers

Recently, two complementary facets within the overall farming systems concept have been identified. Shaner et al. (1982) termed the overall concept Farming Systems Research and Development (FSR&D). The two complementary components of FSR&D that have been recognized (Norman 1982, 1983; Hildebrand and Waugh 1983) were designated by Hildebrand and Waugh as:

- FSIP, the farming systems approach to infrastructural support and policy, and
- FSR/E, the farming systems research and extension approach to technology generation, evaluation, and delivery.

FSIP is more "macro" than is FSR/E. Since it deals with policy and infrastructure, the variables it treats are mainly off-farm and involve more social scientists and economists than agrobiological scientists.

Methodologies frequently include surveys to provide perspective on farming systems that will more accurately predict farmer responses to different policy stimuli. FSIP is applied, farmer-oriented, socioeconomic research, supported by the agrobiological sciences in a team effort. The principal product is information. The primary clients are policymakers and managers of services and infrastructure.

FSR/E is more "micro" in scope and deals mostly with on-farm conditions. Because it is concerned with technology generation, evaluation, and delivery, more agrobiological scientists than socioeconomic scientists are involved, and methodology is concentrated in on-farm biological research with less time devoted to surveys. FSR/E is applied, farmer-oriented, agrobiological research, supported by the socioeconomic sciences in a team effort that includes extension responsibilities. The principal product is technology. The primary clients are farmers.

Traditionally, biological agricultural research has not been closely connected with policy and infrastructure. The systems approach makes it logical and necessary to bring these two components together. If considered in light of the systems approach, the biological agricultural research and the policy and infrastructure are not only compatible but complementary, with each having an input into the other. FSR/E can have significant impact on policy because it can provide more detailed, firsthand information on farms, farmers, and their families than FSIP can obtain easily. Similarly, FSIP can have significant impact on agricultural technology and its dissemination because it can provide FSR/E with more complete information than FSR/E would otherwise be able to obtain. Both components together constitute a more complete agricultural development thrust than either can accomplish singly. The emphasis in this chapter is on FSR/E, which is presently receiving attention as a means of developing technologies for millions of the world's limited-resource farmers.

The systems approach was developed by learning from past experience. This approach is, in itself, a process for learning. It is a process not only for conducting research but also for guiding the research. If this premise is correct, and we strongly believe it is, we can apply the process anywhere and learn what is relevant to specific areas, thus maximizing use of our inventory of technologies and resource endowment around the world.

FSR/E Activities

The systems approach to research and extension is flexible and should be used to fit technologies to the cultural, agricultural, and institutional conditions found in different countries. It will usually involve a sequence of steps similar to the following:

1. Initial agro-socioeconomic characterization and analysis of existing farming systems through close consultation with farmers
   a. Tentative categorizing of farmers into groups who live under similar conditions, follow similar practices, and can be expected to have similar problems
   b. First estimation of problems and constraints
2. Planning and design of first-phase work
   a. Biological research
   b. Continuing agro-socioeconomic characterization
3. Selection, generation, and evaluation of technologies
   a. Commodity and discipline research on experiment stations and in laboratories
   b. Researcher-managed on-farm trials with farmer participation
      —Exploratory trials
      —Site-specific trials
      —Regional agronomic trials
      —Agro-socioeconomic trials
   c. Farmer-managed trials
      —Individual evaluation of acceptability by farmers
      —Refined categorizing of recommendation domains by researchers
      —Initiation of technology transfer activities
4. Information accumulation and analysis
   a. Data from commodity and discipline programs
   b. Agro-technical data from on-farm trials
   c. Economic records on farm enterprises from farmers
   d. Other agro-socio-cultural-economic and political information accumulated through experience and directed surveys of areas
5. Frequently programmed evaluation and planning to:
   a. Refine partitioning of recommendation domains
   b. Make recommendations for mass transfer into specified recommendation domains of acceptable technology
   c. Feedback into the sequential process
   d. Develop plans of work
6. Extension of acceptable results throughout appropriate recommendation domain(s).

In many ways this sequence parallels what farmers have always done. The farmer manages a complex set of biological processes that transform the resources at his or her disposal into useful products, either for home consumption or for sale or trade. The choice of crop and livestock enterprises and the methods and timing of cultivation, husbandry, and harvesting are determined not only by physical and biological constraints, but also by economic, sociopolitical, infrastructural, and policy factors that make up the larger milieu within which the farmer operates.

Within their complex milieu, through a process of trial and error and a number of seasons or generations, farm families move toward appropriate technologies and allocate resources in ways that make best use of those at their disposal given their individual objectives. While the choices available to each farmer are different, those with similar sets of resources and constraints tend to make similar choices as to crops, livestock, and management practices. Those who have responded in similar ways can be grouped together into homogeneous farming systems or domains for recommendation.

FSR/E brings scientific method and additional expertise to bear on problem identification and technology generation. Teams of scientists from different disciplines, working with farmers, can speed up the process and respond more efficiently to a rapidly changing world.

Research/Extension Relationships in FSR/E

> It is now generally accepted that the conceptualization of extension as a service for communicating results of scientific research to farmers neglects the importance of the problem-solving process of education and social organization in agricultural and rural community development (Compton 1984, 79).

The two facets of the systems approach that have received the most attention, and give agricultural research and extension substance as an interdisciplinary activity, are the rapid reconnaissance for quick appraisal of farming areas and the on-farm research.[1]

On-farm research with a systems orientation makes two major contributions to a technological support system. First, it speeds up the development of technology and second, the on-farm research brings farmers and researchers into contact and promotes dialogue between them. In this respect it can be postulated that on-farm researcher/farmer evaluation of technologies speeds up the natural integration of new technologies into the farmer's production system. The result is an extension effect; it is extension even though it was not designed as such (CIAT 1984). If this is true, is not the area of researcher/farmer evaluation of technologies a logical place to strengthen the research/extension link by involving extension in the process? We believe it is.

FSR/E connects extension and fundamental research through an adaptive process of technology development. In fact, FSR/E links extension and research in the problem-solving process and paves the way for "Extension" to play a major role in determining the output from "Research." Extension can

1. Participate in and contribute to the characterization and analysis of farming regions
2. Participate in the supervision of enterprise records and farmer-managed trials for the evaluation of technology
3. Increase the geographic areas covered by evaluation trials and the feedback of information about new technologies to researchers
4. Learn firsthand about the nature and management of new technologies and increase technical capacity, while on the job
5. Initiate transfer of new technologies by using farmer-managed trials in extension activities
6. Receive technical backstopping from on-farm research team members and researchers from the component research programs
7. Use farmer-managed tests as a tool to evaluate new technologies before recommending them to farmers in areas where they have not been tested
8. Provide feedback of infrastructure and policy needs to those researchers and policymakers concerned with FSIP

If extension gains materially from collaborating with research, through information and more relevant research results, research gains even more through expanded testing and evaluation of new technologies at the farm level, and by the increased demand for relevant technology created by extension and by farmers.

In the final analysis, the best instrument for augmenting research and extension ties is the development of proven and useful technology. Extension in developing countries suffers from a poor image and lack of technical competence. Extension participation in the evaluation of technology can improve the impact of technology, and thus enhance the image of both research and extension.

Implementation of FSR/E

Implementation of FSR/E may imply change in organization, in strategies, and in methodologies. Are these changes drastic ones or just new dimensions and shifts that are not highly disruptive? In those cases where research and extension are now poorly organized, major changes very likely will be required. Although the systems approach does not necessitate major changes, changes will very likely be needed in at least three areas: (1) the philosophy, strategy, and methodologies; (2) the development of interdisciplinary activities, which can only be done by including these disciplines within the research/extension organization(s); and (3) the establishment of on-farm research teams.

Structure and Organization

Success or failure of FSR/E will probably depend on the quality of the institution(s) in which it is embodied and the surrounding institutions from which support and collaboration are needed. Since FSR/E is more dynamic, more dispersed in its operations, integrates more disciplines, and requires more collaboration from other institutions, it may suffer from poor institutional management and administrative services more than station-based research. These deficiencies are related more to style and quality of governmental institutions than to specific organization, and we feel that the systems approach can be implemented under different institutional organization.

In cases where research is station-based and commodity- and discipline-oriented, the authority for control of budgets and physical resources, and even supervision of research groups, is sometimes placed with directors of a few major experiment stations who dominate the pattern of research. Such stations may be nuclei of resistance to change; and, if used as headquarters for field operations, they may foil collaboration—especially between commodity research and on-farm teams. This is an example of when reorientation of research may be important in order to increase participation of on-farm research teams in the overall research/extension process.

Components of Organization and Structure for FSR/E

Important concerns for determining organizational structure include the nature and customs of governmental institutions; physical, financial and human resources; areas to be covered; and the status of agriculture in the country (Waugh et al. 1983). Institutional collaboration is another concern, especially in the case of extension, which requires the dedication of time and resources, especially to the evaluation of technologies in collaboration with the on-farm area research team. We believe that, regardless of the organizational structure, a research/extension organization (or organizations) capable of applying a farming systems approach will have at least the following six functional groups:

1. *Top management and the management team.* Management is usually visualized as the function and responsibility of a few top officials of an organization. The systems approach, however, requires a multidisciplinary participation in management decisions. A management team, headed by top management, including heads of programs and units of organization, can be developed. When top management wishes to communicate with lower levels in the hierarchy it can be done through members of the management team. Owing to the nature of the systems approach, which requires linkages and coordinated action, the management team takes on greater importance than in the case of most traditional research/extension programs.

2. *Component and discipline research.* Component and discipline research with a farming systems approach changes somewhat as compared to traditional research. The commodity and biological disciplines direct their work to meeting the needs of specific farming areas where on-farm research teams and extension agents are evaluating, integrating, and disseminating technologies. Since most countries cannot afford, and do not need, more than one commodity team for a given commodity, the commodity teams have national responsibilities and work with several on-farm research teams. Much of their work is still station-based, but includes off-station work as needed.

The socioeconomic team (component research) also has national responsibility, even though some members might have responsibility for specific areas or regions. Most of their work is not station-based, however, but is conducted in the specific farming areas in collaboration with on-farm teams.

3. *On-farm research teams assigned to specific farming areas.* Traditional station-based researchers usually conduct regional trials distributed over a range of ecological zones. Some of these may be on-farm trials, an important methodology for the generation of new technologies, especially new crop varieties. Commodity teams usually focus on a single crop, however, and conduct relatively few and widely separated trials under relatively optimal conditions that do not reflect real farm conditions. On-farm research teams assigned to specific

farming areas generally have not been used in traditional research.

On-farm research teams (a) are assigned to specific, delimited farming areas and are responsible for the integration of technologies into the local farming systems, (b) study alternatives at the local level, (c) evaluate technologies at the local level under real farm conditions and with the involvement of farmers, and (d) work in collaboration with extension in the evaluation of technologies and function as a technical backstop for extension activities.

The size of the on-farm team varies according to the geographic area to be covered, available transportation, infrastructure, and farming conditions. However, the work needed to establish the basic parameters of farming technology such as response to fertilizer, weed and pest control, time of planting, plant population, the details of farmer practices, variability of soil fertility, and the qualities and characteristics of technology acceptable to farmers constitutes a basic load of work that cannot be efficiently carried out without adequate numbers of qualified agronomic personnel.[2] Team composition and size vary considerably, but the main work, after the original planning, is biologic research. The team participates in an information-gathering and learning process to update the first characterization and develop technology for the area. Social scientists continue close contact with the team, but their work load is now less, compared to that of agronomists and/or animal scientists. Enterprise records (Hildebrand 1979) are another valuable tool to be applied in specific farming areas; however, it is usually not feasible to assign a full-time professional economist to an on-farm team. The agronomists, for example, can gather economic data with some help and supervision of an economist. In this manner, one economist can work with two, three, or even four on-farm research teams. The situation is similar for other social science disciplines.

4. *Extension agents assigned to specific farming areas.* To be effective, the link between research and extension must occur at the farming area level. Links that are agreed upon in the central offices between research and extension do not automatically bring about interchange and dialogue in the field and do not take advantage of what is learned from and about the farmers. On-farm teams, supported by commodity and other discipline personnel, effectively become "subject matter specialists" with whom extension collaborates in the evaluation of technology. On-farm teams become technological support units for extension. Extension personnel thus gain first-hand experience with technologies that strengthen the basis for dissemination to farmers as well as encourage feedback from extension to research.

5. *Administrative support services.* Every organization needs administration for handling funds, purchasing, financial and personnel records, contracts, etc. Bureaucratic organizations must execute administrative procedures according to rules and regulations of government. As a result, the administrative procedures are frequently managed to meet the requirements of government rather than to support technical programs. Rules and regulations cannot be ignored, but the responsibility of administration should be to meet the requirements of the technical programs. This usually can be done if the administration truly has as its objective the support of technical programs and if it understands their needs.

6. *Technical support units for research.* Technical support units, such as experiment stations, laboratories, warehousing, and library and editorial services, are needed for the systems approach as much as they are needed for traditional research. To keep overhead costs to a minimum, such support should be developed according to specific needs, rather than on a generic basis that says "a soils laboratory is needed to conduct soils research." There may be several means of developing support for the research units through the involvement of groups such as the IARCs, universities, and regional organizations, in which case the research organizations may not need to develop support internally. In fact, they may be duplicating costly services and support.

Group or Unit Interrelationships

The development of interrelationships among and between extension and research personnel, component research and on-farm teams, and biological and social scientists is a key factor for implementation. In some cases FSR/E has been annexed to research organizations rather than used as an integrated approach for all research. In such cases the diagnosis of farming areas, the on-farm research, and the review of results and planning probably has had little relation with commodity and other component research. In our opinion, FSR/E cannot be a replacement or substitute for component research nor conducted separately from it. It is a new dimension, which should be coordinated with component research.

Another pattern of implementation places the responsibility of the farm-level research work on the experiment station-based personnel who work in commodity-oriented programs. This does not seem to be the most effective pattern because it dilutes the commodity effort while continuing a monocrop focus. Neither does it encourage the development of on-farm teams who are dedicated to and become knowledgeable about specific farming areas. While we find no specific documentation that examines and substantiates this observation, we feel that teams of dedicated people who have continuous dialogue with farmers and their families and constant contact with the farming problems of a specific and defined geographic area are more successful in developing relevant technologies than teams that combine work on experiment stations with periodic travel.

These on-farm teams do not commonly exist in traditional research institutions. Their performance can be critical for the success or failure of the systems approach. They have the tremendous task of integrating complicated component technology into farming systems in a manner that is relevant to the prevailing conditions. To do this they must help orient commodity and discipline teams. Their performance depends upon dialogue and relations with farmers and their families, which requires comprehension of a broad gamut of social/economic factors. They must have links with extension personnel and support them.

While there are many constraints to developing the systems approach, problems associated with establishing the on-farm activity is one of the greatest.

Limited experience suggests some actions to promote coordinated activity among disciplines:

1. Establish an equitable institutional "home" for all disciplines; for example, organize the social scientists in an identifiable institutional unit, with a director and an assigned budget in a manner similar to that for the biological scientists. Remove, insofar as possible, any stigma associated with "second class citizenship" which may exist when, for example, extension is viewed as inferior to research. Consider moving outstanding and prestigious personnel to the group that is viewed as inferior.
2. Encourage cooperation, dialogue, and joint action, but with designated leadership and defined roles and responsibilities.
3. Organize periodic review and planning with joint participation of all disciplines.
4. Manage technical activities without unilateral decisionmaking by one group or discipline.
5. Make all work purposeful in terms of what is beneficial to the clientele, and not solely a contribution to science, a contribution to a particular discipline, or for the satisfaction of personal curiosity.

Constraints to Implementation

Diversity of opinion. Since the farming systems approach is based on the premise that biological, physical, social, and economic disciplines can produce more relevant technology as a coordinated effort than they can separately, it is important to have a consensus of objectives and strategies, and to some degree the methodologies. However, the process of defining team objectives and strategy involves more than simply coordinating four discipline groups. Multidisciplinary action should produce an interdisciplinary result.

Professional groups. General lack of understanding of the farming systems approach may be cause for resistance, especially if it is viewed as a substitution for current programs rather than as a new dimension favorable to current programs. Since research personnel have been trained in commodities and disciplines they may have little interest in understanding cultural-economic aspects of farming and see little reason for change.

Government and its institutions. Many institutions of government have characteristics that constitute constraints to the implementation of a farming systems approach (Norman 1983). These characteristics, some of which are mentioned below, fortunately are not part of all institutional environments. But when they are present, it is important to recognize them and learn how to live with them or how to overcome them. Otherwise the systems approach may be little or no better than the present commodity-oriented programs.

Bureaucratic systems of government may be more interested in perpetuating bureaucracy than instituting changes to serve clientele. Bureaucratic systems have established organizational structures and procedures and are highly resistant to change. Some bureaucracies may acquiesce to change, then change only names of institutions and titles of positions so that no new concepts are truly institutionalized.

Organizational structure may have placed much of the power of leadership and control of budgets and physical resources at major experiment stations that dominate the pattern of research. These stations may not only resist change but foil collaboration even when such would further their commodity and discipline research.

Even though there is consensus for change, the institutional capacity for developing adequate logistic and administrative support may be lacking.

The general economic condition of the country, even with outside contributions, may cause low salaries and rapid turnover of personnel. And even though financial support is adequate to support a modest but high-quality program, the program may suffer because government insists on too great a coverage, which dilutes efforts.

These characteristics are listed to emphasize that if a systems approach is to be implemented successfully, there must be an assessment in each instance to discern problems and devise approaches that will overcome the constraints (Waugh et al. 1983; Norman 1983).

Periodic Review and Planning

Periodic review and planning are central to the systems approach. This means updating, analyzing, and interpreting information and developing recommendations and plans of work. This process requires direction and management; it is a management mechanism. It is a structured and organized activity through which management articulates external and internal policy and evaluates the quality of personnel and their work. It is a mechanism for presenting technical results that can be selected for inclusion in the annual report.

Review and planning utilize all available biological and socioeconomic information. This implies that analysis and interpretation of data are undertaken rapidly and efficiently. Complicated design and analysis that cannot be completed in time for review and planning contribute little. Results of the most recent research, then, should serve as the basis for the plan of work for the next series. Since FSR/E involves a wider range of disciplines and more diverse groups than traditional research, coordination is more important than for more traditional research approaches.

Future Potential

In the United States

As a result of the technological revolution, government-supported research in the United States has drifted more and more toward specialization. It is discipline-driven research that meets the demands of a sophisticated and well-educated clientele, many of whom live on farms and are full-time professional managers. Commercial agricultural development firms also cater to this clientele. Research findings are beneficial to commercial agriculture and can be readily applied by this client group. Ruttan (1983) has called to our attention the impact of technology by comparing maize yields in Argentina with yields in the United States. Thirty years ago the yields in both countries were about the same. Today U.S. yields are more than twice as high as those in Argentina. At the same time, however, smaller farms in the

United States, many with part-time management, were benefiting less from the technological revolution because they were unable to adapt much of the commercial technology. A void was created in the supply of technology for smaller and more-limited-resource farms. This was due in part to the fact that technology development was left in the hands of commercial firms who were attracted to early profits with large-scale operations, and in part to the drift to fundamental research by public organizations that began to lose sight of the need for technology adaptation and development.

Small farms in the United States also have special infrastructural and policy requirements that must be better identified in the future. Improved links between research and extension as a result of farmer involvement in technology development will strengthen the land-grant system and the agricultural sector for both small and large farms. This fine-tunes the land-grant philosophy to help correct institutional drift toward specialization.

The farming systems, client-participatory approach to improve research and extension efficiency has great potential for providing technology development for more-limited-resource farmers. Although the farming systems approach has been utilized over the past few years primarily in the development of technology for small-scale, limited-resource farmers, it can also be effective in developing technology for commercial agriculture. The approach can have a favorable effect on the efficiency of agricultural research for clients at all levels of economic activity. Coordinated research undertaken over a broad range of agricultural and socioeconomic environments provides fundamental researchers more precise information on needs and desires of different clientele groups.

Furthermore, client participation in technology development augments limited research resources. In limited-resource research and extension institutions this is, and will continue to be, an important consideration.

In Developing Countries

Because many research and extension institutions in developing countries were modeled after the land-grant system, similar problems and opportunities apply to them but are magnified many times over those of the the United States, owing to an even greater scarcity of resources. Furthermore, in most developing countries the proportion of farmers who receive little benefit from established research and extension institutions is much greater. This magnifies the problem further. Hence, if the farming systems approach to research and extension has great potential for serving farm clientele in the United States, its potential is even greater in developing countries. Because much agricultural technology is effective when moving from small scale to large scale, but not "scale neutral" in moving from large to small scale, it behooves developing countries to concentrate government research and extension efforts on the small, limited-resource farmer.

The farming systems philosophy is composed of several forward-looking principles to influence the future of agricultural research and extension. It promotes greater interdisciplinary activity in the approach to understanding farmers' problems and fewer unilateral decisions by discipline and commodity researchers. Foremost is the principle of participation by the farmer and those who work closely with the farmer to identify problems and solutions. Let us state these principles as conditions: (1) *Mutual trust and respect.* Trust generates confidence and removes a barrier to farmer involvement. (2) *Involvement.* Problem identification is at the level where the facts are most obvious. Farmers support what they help create. (3) *Communication.* Farmers want research and extension people to know as much as possible about their work environment and their problems. With this knowledge, technology is more readily acceptable and risk is reduced for the farmer and better understood by research and extension. Farmers then have higher expectations for their own well-being from research and extension. (4) *Higher expectations.* Honoring higher expectations for survival and advancement is a basis for achieving farmer and institutional goals.

This philosophy is rooted in the U.S. land-grant system and the refinements implied by the relatively new term, "farming systems." The philosophy can stimulate the development of concepts, policies, institutional structures, and administrative procedures to achieve agricultural success. The future holds some significant adjustments for agricultural research and extension. But in development, attention to structure alone may not be conducive to success unless we understand and contribute to the philosophical underpinnings of these concepts and structures. The rise or fall of the term "farming systems" is not in itself important, but the concept will profoundly influence agricultural research and extension worldwide.

The drift of U.S. universities toward fundamental research develops scientific knowledge but not applicable technology from this knowledge and leaves the universities unprepared to maximize the systems approach for development. The universities have an outstanding scientific capability but a paucity of personnel who can deal with foreign governments to institutionalize the systems approach.

This arises from two situations: (1) The lack of steady government support and participation in agricultural development and (2) the system of staff evaluation within universities that does not allow full-time dedication to development work because of the requirement for individual contributions to scientific knowledge. The development of applicable technologies is thus neglected.

The future will demand not only strengthened farm policy but also policy for research and development where farmer attitudes will be more influential. Technology development involving farmers is more political than traditional research; and thus policy interpretation and articulation are more important. The researcher on the experiment station operates in an apolitical situation, isolated from clientele evaluation. Research on farms is viewed in the light of public needs; when researchers are successful or make mistakes they are more visible.

Farmers are more than half of the world's population. Any criterion that is to relate to their well-being, in balance with the urban half of the world, should improve agricultural development on a broader scale and provide greater food and total security for all. Should this criterion be put to use, the door will open for a transformation in international agricultural research and development.

A New Affirmation

One of the most important aspects of the farming systems efforts in the 1980s was the creation of the Annual Farming Systems Symposium. Originally established by Kansas State University, the eighth annual symposium was held at the University of Arkansas in October 1988. The symposium attracts farming systems practitioners and administrators from the United States and around the world. International participants come from projects supported by USAID, the International Development Research Centre of Canada, the World Bank, the Inter-American Development Bank and other major donors. The International Agricultural Research Centers, who jointly issued a statement confirming the importance of the farming systems perspective in February, 1986, are also always represented. A representative of the farming systems Society of the People's Republic of China was expected to attend in 1988. Many practitioners and administrators from U.S. universities with domestic or international farming systems activities are also always present. Besides the symposium proceedings, a rapidly increasing number of books on farming systems research and extension and professional journals that publish articles on farming systems research have created an expanding source of available literature on FSR/E.

At the time this chapter was written, the Farming Systems Support Project (FSSP) funded by USAID's Office of Science and Technology and its Office of Agriculture was in operation. Its mandate was to provide support to existing USAID-funded farming systems projects and to USAID missions developing new farming systems projects in their host countries.

FSSP had activities in many countries around the world. It also produced a newsletter which was printed in three languages and distributed to about 5,000 people. In 1983, in response to a request run in the *FSSP Newsletter*, 260 farming systems projects were identified in more than seventy-five countries. Not all these projects were supported by USAID.

Twenty-one U.S. universities and four private organizations constituted the FSSP, and the FSSP was coordinated by the University of Florida. Funding for the FSSP terminated at the end of 1987. However, the network created by that effort continues, and modest funding is still being provided by USAID. The network has created a steering committee that has been asked by USAID to study alternatives for future activities and sources of funding.

In summary, farming systems research and extension is becoming firmly established in many parts of the world. The name may vary from country to country, but the principles upon which the activities are based are the same. Emphasis is shifting to include more infrastructure and policy concerns as the methodology for technology development is more firmly established and understood. Persons with farming systems experience are beginning to appear in the administrative ranks of research and extension institutions in developed and developing countries. The result will be research and extension organizations that are more responsive to clientele needs and a more efficient process for the development of agriculture throughout the world.

Notes

1. While rapid reconnaissance and on-farm research are two important parts of FSR/E, these two by themselves do not constitute a systems approach. They must be integrated with commodity and discipline research and extension, all of which in turn must be coordinated with policy and support services.

2. Average on-farm area research team size in ICTA in Guatemala is about 5.5 persons, mostly professional agronomists with fewer than ten years of experience. A few are "Peritos Agrónomos," a vocational

secondary education degree. Team members tend to "specialize." For example, one might be inclined to develop expertise on soils, another on weed control, another on statistical analysis, and another on economics. In this manner one team member complements another. All team members generally conduct the same kinds of trials, however, and the "specialization" is not reflected in the assignment of responsibility for conducting field work (Waugh 1980b).

References

Abalu, G. "Organization and Management of Farming Systems Research: Experiences from Nigeria." Paper prepared for the ISNAR/CIMMYT Workshop on Issues in Organization and Management of Research with a Farming Systems Perspective, The Hague, September 27-30, 1983.

Barr, T. N. "The World Food Situation and Global Grain Prospects." *Science* 214(1981):1087-95.

Bradfield, R. "Toward More and Better Food for the Filipino People and More Income for Her Farmers." New York: Agricultural Development Council, December 1966.

CIAT. "It's a Good Bean." *CIAT International* 3: no. 2, (1984). Centro Internacional de Agricultura Tropical, Cali, Colombia.

CIMMYT. "The Puebla Project, Seven Years of Experience: 1967-73." Centro Internacional de Mejoramiento de Maíz y Trigo, Mexico City, 1974.

Compton, J. Lin. "Linking Scientist and Farmer: Re-thinking Extension's Role." In *World Food Issues,* M. Drosdoff, ed. Ithaca: Cornell University Program in International Agriculture, 1984, 79-84.

Gilbert, E. H., D. W. Norman, and Fred E. Winch. *Farming Systems Research: A Critical Appraisal.* Rural Development Paper No. 6, East Lansing, Department of Agricultural Economics, Michigan State University 1980.

Harwood, Richard R. *Small Farm Development: Understanding and Improving Farming Systems in the Humid Tropics.* Boulder, Colo.: Westview Press, 1979.

Hildebrand, P. E. "The ICTA Farm Record Project with Small Farmers." Instituto de Ciencia y Tecnología Agricolas, Guatemala City, 1979.

Hildebrand, Peter E., and Robert K. Waugh. "Farming Systems Research and Development." *Farming Systems Support Project Newsletter* 1, (1983). Gainesville, Fla.: University of Florida.

Norman, David W. "The Farming Systems Approach to Research." *In FSR Background Papers. Paper No. 3,* Farming Systems Research Paper Series. Manhattan, Kansas: Kansas State University, 1982

——. *Some Problems in the Implementation of Agricultural Research Projects with a Farming Systems Perspective.* Farming Systems Support Project, Networking Paper No.3. Gainesville, Fla.: International Programs, Institute for Food and Agriculture Sciences, University of Florida, 1983.

Rice, E.B. *Extension in the Andes: An Evaluation of Official U.S. Assistance to Agricultural Extension Services in Central and South America.* Cambridge: MIT Press, 1974.

Ruttan, Vernon W. "The Global Agricultural Support System." *Science* 222

(1983):4619.

——. "Integrated Rural Development Programs: A Historical Perspective." Unpublished manuscript, 1984.

Shaner, W. W., P. F. Philipp, and W. R. Schmehl. *Farming Systems Research and Development*. Boulder, Colo.: Westview Press, 1982.

Tarte, R. "Experience in Organization and Management of Research with a Farming Systems Perspective in Panama." Presented at the ISNAR/CIMMYT Workshop on Issues in Organization and Management of Research with a Farming Systems Perspective, The Hague, September 17-20, 1983.

Waugh, Robert K. "ICTA: Four Years of History." Instituto de Ciencia y Tecnología Agricolas, Guatemala City, 1975.

——. "La investigación agricola en el áreas de influencia del PCCMCA y su proyección hacia el futuro." Annual meeting of the PCCMCA, Guatemala City, 1980a.

——. "Staffing and Funding in the Instituto de Ciencia y Tecnología Agricolas (ICTA)." Farming Systems Research Symposium, sponsored by OICD, USDA. Washington, D.C., December 8-9, 1980b.

Waugh, Robert K., J. Meiman, and James K. McDermott. *Institutional Assessment for Implementing a Systems Support Approach to Agricultural Research and Extension.* Farming Systems Support Project Working Paper No. 101. Gainesville, Florida: International Programs, Institute of Food and Agricultural Sciences, University of Florida, 1983.

Whyte, William F. *Participatory Approaches to Agricultural Research and Develpment: A State-of-the-Art Paper.* Special Series on Agricultural Research and Extension, No. 1. Ithaca, New York: Rural Development Committee, Center for International Studies, Cornell University, 1981.

Zandstra, Hubert, Kenneth Swanberg, Carlos Zulberti, and Barry Nestel. *Caqueza: Living Rural Development.* Ottawa: International Development Research Centre, 1979.

Index

About the Book

Addressing the transformation of the nature and functions of agricultural research and development institutions and their programs, this book explores the major changes and identifies key trends and issues. The authors also document the influence on agricultural development of such U.S. initiatives as experiment stations and assess the effectiveness of various research and education strategies in Asia, Africa, Latin America, and the Middle East.

The topics discussed, all central to the transformation process, include the simplified communication of scientific knowledge, the role of women and the linking of indigenous and scientific agricultural knowledge. The final chapters consider the functioning of the international agricultural research network and the development of farming systems research and extension strategies.

J. Lin Compton is associate professor of extension education, international agriculture, and Asian studies at the University of Wisconsin-Madison. He has worked for more than ten years in a variety of agricultural and rural community development efforts in several Asian countries, and in 1972-1974 he was head of the Education and Culture Department of the International Institute of Rural Reconstruction in the Philippines.